INTERNATIONAL WILDLIFE ENCYCLOPEDIA

THIRD EDITION

Volume 14

PAR–POO

Marshall Cavendish Corporation
99 White Plains Road
Tarrytown, New York 10591–9001

Website: www.marshallcavendish.com

Library of Congress Cataloging-in-Publication Data

Burton, Maurice, 1898-
    International wildlife encyclopedia / [Maurice Burton, Robert Burton] .-- 3rd ed.
        p. cm.
    Includes bibliographical references (p.).
    Contents: v. 1. Aardvark - barnacle goose -- v. 2. Barn owl - brow-antlered deer -- v. 3. Brown bear - cheetah -- v. 4. Chickaree - crabs -- v. 5. Crab spider - ducks and geese -- v. 6. Dugong - flounder -- v. 7. Flowerpecker - golden mole -- v. 8. Golden oriole - hartebeest -- v. 9. Harvesting ant - jackal -- v. 10. Jackdaw - lemur -- v. 11. Leopard - marten -- v. 12. Martial eagle - needlefish -- v. 13. Newt - paradise fish -- v. 14. Paradoxical frog - poorwill -- v. 15. Porbeagle - rice rat -- v. 16. Rifleman - sea slug -- v. 17. Sea snake - sole -- v. 18. Solenodon - swan -- v. 19. Sweetfish - tree snake -- v. 20. Tree squirrel - water spider -- v. 21. Water vole - zorille -- v. 22. Index volume.
    ISBN 0-7614-7266-5 (set) -- ISBN 0-7614-7267-3 (v. 1) -- ISBN 0-7614-7268-1 (v. 2) -- ISBN 0-7614-7269-X (v. 3) -- ISBN 0-7614-7270-3 (v. 4) -- ISBN 0-7614-7271-1 (v. 5) -- ISBN 0-7614-7272-X (v. 6) -- ISBN 0-7614-7273-8 (v. 7) -- ISBN 0-7614-7274-6 (v. 8) -- ISBN 0-7614-7275-4 (v. 9) -- ISBN 0-7614-7276-2 (v. 10) -- ISBN 0-7614-7277-0 (v. 11) -- ISBN 0-7614-7278-9 (v. 12) -- ISBN 0-7614-7279-7 (v. 13) -- ISBN 0-7614-7280-0 (v. 14) -- ISBN 0-7614-7281-9 (v. 15) -- ISBN 0-7614-7282-7 (v. 16) -- ISBN 0-7614-7283-5 (v. 17) -- ISBN 0-7614-7284-3 (v. 18) -- ISBN 0-7614-7285-1 (v. 19) -- ISBN 0-7614-7286-X (v. 20) -- ISBN 0-7614-7287-8 (v. 21) -- ISBN 0-7614-7288-6 (v. 22)
    1. Zoology -- Dictionaries.   I. Burton, Robert, 1941- . II. Title.

    QL9 .B796 2002
    590'.3--dc21
                                                      2001017458

Printed in Malaysia
Bound in the United States of America

07 06 05 04 03 02   8 7 6 5 4 3 2

Brown Partworks
Project editor: Ben Hoare
Associate editors: Lesley Campbell-Wright, Rob Dimery, Robert Houston, Jane Lanigan, Sally McFall, Chris Marshall, Paul Thompson, Matthew D. S. Turner
Managing editor: Tim Cooke
Designer: Paul Griffin
Picture researchers: Brenda Clynch, Becky Cox
Illustrators: Ian Lycett, Catherine Ward
Indexer: Kay Ollerenshaw

Marshall Cavendish Corporation
Editorial director: Paul Bernabeo

Authors and Consultants

Dr. Roger Avery, BSc, PhD (University of Bristol)

Rob Cave, BA (University of Plymouth)

Fergus Collins, BA (University of Liverpool)

Dr. Julia J. Day, BSc (University of Bristol), PhD (University of London)

Tom Day, BA, MA (University of Cambridge), MSc (University of Southampton)

Bridget Giles, BA (University of London)

Leon Gray, BSc (University of London)

Tim Harris, BSc (University of Reading)

Richard Hoey, BSc, MPhil (University of Manchester), MSc (University of London)

Dr. Terry J. Holt, BSc, PhD (University of Liverpool)

Dr. Robert D. Houston, BA, MA (University of Oxford), PhD (University of Bristol)

Steve Hurley, BSc (University of London), MRes (University of York)

Tom Jackson, BSc (University of Bristol)

E. Vicky Jenkins, BSc (University of Edinburgh), MSc (University of Aberdeen)

Dr. Jamie McDonald, BSc (University of York), PhD (University of Birmingham)

Dr. Robbie A. McDonald, BSc (University of St. Andrews), PhD (University of Bristol)

Dr. James W. R. Martin, BSc (University of Leeds), PhD (University of Bristol)

Dr. Tabetha Newman, BSc, PhD (University of Bristol)

Dr. J. Pimenta, BSc (University of London), PhD (University of Bristol)

Dr. Kieren Pitts, BSc, MSc (University of Exeter), PhD (University of Bristol)

Dr. Stephen J. Rossiter, BSc (University of Sussex), PhD (University of Bristol)

Dr. Sugoto Roy, PhD (University of Bristol)

Dr. Adrian Seymour, BSc, PhD (University of Bristol)

Dr. Salma H. A. Shalla, BSc, MSc, PhD (Suez Canal University, Egypt)

Dr. S. Stefanni, PhD (University of Bristol)

Steve Swaby, BA (University of Exeter)

Matthew D. S. Turner, BA (University of Loughborough), FZSL (Fellow of the Zoological Society of London)

Alastair Ward, BSc (University of Glasgow), MRes (University of York)

Dr. Michael J. Weedon, BSc, MSc, PhD (University of Bristol)

Alwyne Wheeler, former Head of the Fish Section, Natural History Museum, London

Contents

PARADOXICAL FROG

THE PARADOXICAL FROG looks like any ordinary frog in outward appearance, but it is quite the reverse of most other frogs because at metamorphosis it becomes several times smaller than its tadpole. The adult frog is up to about 3 inches (7.5 cm) long. Its hind feet are webbed, but the toes project beyond the webbing more than usual. It has the green-brown coloration typical of frogs, with darker spots and blotches, and the hind legs have a harlequin coloring of yellow and black.

There are several subspecies of the paradoxical frog, the best known living on the Caribbean island of Trinidad, in the northeastern area of South America and in part of the Amazon basin. Others are found in other parts of South America as far south as northern Argentina.

Heard but not seen

Paradoxical frogs can often be heard making coughing grunts, almost like those of pigs. The frogs are rarely seen, however. They seldom come onto land and whenever they come to the surface of the water they are very hard to spot. Paradoxical frogs usually expose just their eyes or nostrils, generally among small water plants crowding the surface. The moment the frogs are disturbed they dive. They have an added protection in that their skin is unusually slippery, so they are extremely difficult to hold.

Another peculiarity of paradoxical frogs is that the toes have an extra joint, which gives them greater length. This characteristic is linked with the frogs' method of feeding. They use their toes to stir up the sediment at the bottom of shallow lakes to find the small mud-dwelling invertebrates that they eat. The first toe on each forefoot is opposable to the others and is used as a thumb for grasping food.

From giant to dwarf

The outstanding feature of the paradoxical frog is the size of the tadpole, which may be 9 inches (23 cm) long. When the tadpole changes into a froglet, it shrinks to about 1¼ inches (3 cm) long. This turn of events is so unusual that the first scientists to study these frogs could not believe they belonged to the same species as the tadpoles. When a tadpole shrinks to this degree, all the internal organs must shrink in proportion,

The paradoxical frog (below) belongs to the family Pseudidae along with Pseudis minuta and the two members of the genus Lysapsus.

PARADOXICAL FROG

CLASS	**Amphibia**
ORDER	**Anura**
FAMILY	**Pseudidae**
GENUS AND SPECIES	***Pseudis paradoxa***

LENGTH
Adult: up to 3 in. (7.5 cm); tadpole: up to 9 in. (23 cm)

DISTINCTIVE FEATURES
Adult: protuberant eyes; strong, robust hind limbs; fully webbed feet; long toes that project unusually far from webbing; tadpole: enormous size

DIET
Pond-dwelling invertebrates

BREEDING
Breeding season: September–May; number of eggs: about 200; hatching period: about 40 days; tadpole reaches maximum length of 9 in. (23 cm) at 4 months; metamorphoses into much smaller adult frog that is 1¼ in. (3 cm) long at first

LIFE SPAN
Not known

HABITAT
Large swamps and large, still bodies of water; some subspecies in arid regions spend dry season in mud at bottom of ponds

DISTRIBUTION
Tropical South America, east of Andes

STATUS
Not threatened

Paradoxical frog

a time there were several different scientific names for the different stages in the paradoxical frog's life history. This is not the only time that such a misunderstanding has occurred, but it is rare for there to be such a great reduction in size from the young or larval animal to the adult. It raises many questions to which scientists do not yet know the answers. For example, how large are the paradoxical frog's eggs and do they contain large quantities of yolk to feed the growing tadpole in its early stages? Another unanswered question is what advantage can there be in a tadpole growing so large and then shrinking so much as it turns into a froglet?

Tadpoles to market

One reason information on the paradoxical frog and its relatives is so scanty is that these frogs are difficult to find and therefore to keep in captivity in sufficient numbers to study them. Three specimens of one species were collected 100 years ago, but there is no record of its having been seen since. In the last decades of the 20th century, however, South American zoologists found more specimens of other species in areas south of the Amazon basin, so more information has begun to trickle through. The tadpoles are easier to catch as they can be netted, but even this operation apparently is not simple. The local people manage, though. They catch the tadpoles, as well as the adults, on hooks baited with grasshoppers and then sell them in the markets.

The paradoxical frog has extra-long toes. This characteristic helps the animal to stir up the mud when it is feeding and may also help it in its aquatic life by increasing the area of webbing on its feet.

a process that was quite unheard of at that time. As a result the adult frog was given one scientific name and the tadpole was given another. Only when the change from tadpole to frog had been observed was it realized that a single species was involved. The situation was so puzzling that for

PARAKEET

Although it is common in some areas, the eastern slaty-headed parakeet, Psittacula himalayana finschii (above), is becoming scarce over wide areas of its Southeast Asian range.

PARAKEETS ARE MEMBERS of the parrot family, the best-known member of which is the budgerigar, *Melopsittacus undulatus* (described elsewhere in this encyclopedia). It is very difficult to give a definition of the name parakeet; practically every book on birds, and certainly all dictionaries, differ in their views. In general, parakeets can be defined as small, brightly colored parrots with long tails, but the name has been given to many genera and species in tropical America as well as in southern and Southeast Asia and Australasia. Parakeets are divided into three tribes, or subfamilies, of the parrot family Psittacidae: the Platycercini, Psittaculini and Arini. Between them, these tribes contain 251 parakeet species, although not all of the birds have the word *parakeet* as part of their common name.

New World parakeets

Identification in this large group of small, colorful parrots is especially confusing because they are not all closely related. The American parakeets, or conures, which belong to the tribe Arini, range from Mexico to Paraguay. They are related to the macaws, but differ from them in that the lores (the space on each side of the head between the bill and the eyes and cheeks) are feathered and they have a large swollen bill, whereas that of macaws is compressed. Also, the fourth feather of the wing is long and narrow, and the nostrils are exposed. These parakeets are usually a shade of green, yellow or orange, with blue on the wings and sometimes red on the head and breast. The slender-billed parakeet, *Enicognathus leptorynchus*, is another American species. It is 15 inches (37.5 cm) long and has pale green plumage with crimson on the forehead and around the eyes and lores, and it has a faint patch of red on the belly. It lives in Chile, in large flocks numbering hundreds of birds. Although it keeps mainly to the forests, the slender-billed parakeet comes out to attack crops from October to April, feeding on cereal crops and the roots of grass. The gray-breasted parakeet, *Myiopsitta monachus*, is green with gray on the head, throat and breast. It is nearly 12 inches (30 cm) long and ranges from Bolivia south to Argentina. There are several related species in South America. The smaller American species include the green parakeet, *Brotogeris tirica*, 7–10 inches (17.5–25 cm) long, which is found in northern South America and Central America.

Old World parakeets

The Old World parakeets, tribe Psittaculini, are more closely related to the true parrots. Like their New World counterparts, they are also mainly green. The rose-ringed parakeet, *Psittacula krameri*, of Africa and Asia, which grows up to 18 inches (45 cm) long, has a graduated tail in which the two central feathers are long and narrow. It has a notch in the upper half of the bill and a narrow rose-colored collar. The Indian blossom-headed parakeet, *P. cyanocephala*, a near relative, is more striking still. The head of the male is tinged with red and it also has plum-colored cheeks and a thin black collar.

Australasian parakeets

The parakeets native to Australasia belong to the tribe Platycercini. The grass parakeets, of which the torquoisine, *Neophema pulchella*, is the best known, live in southern Australia and Tasmania.

ROSE-RINGED PARAKEET

CLASS	**Aves**
ORDER	**Psittaciformes**
FAMILY	**Psittacidae**
GENUS AND SPECIES	***Psittacula krameri***

ALTERNATIVE NAME
Ring-necked parakeet

WEIGHT
3⅓–5 oz. (95–145 g)

LENGTH
**Head to tail: 14½–17 in. (37–43 cm);
wingspan: 16½–19 in. (42–48 cm)**

DISTINCTIVE FEATURES
**Male: pale yellow-green plumage; black chin;
narrow black and rose-pink collar on side of
neck; blue nape; very long tail with blue-green
central feathers; dark red upper mandible;
black lower mandible. Female: lacks black,
rose and blue on head and neck.**

DIET
Fruits, seeds, nuts, flower petals and nectar

BREEDING
**Age at first breeding: 1 year; breeding
season: December–May (West Africa),
January–May (Asia); number of eggs: 3 or 4;
incubation period: about 22 days; fledging
period: 49 days; breeding interval: 1 year**

LIFE SPAN
Up to 20 years

HABITAT
**Mostly deciduous habitats, from semidesert
to secondary forest; also large gardens and
agricultural land with scattered trees**

DISTRIBUTION
**Senegal east to Somalia; Pakistan, India and
Myanmar (Burma) east to southwestern
China; introduced to southeastern England**

STATUS
Common or abundant

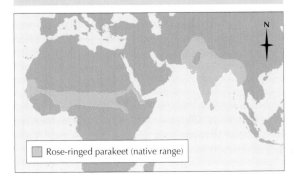

Rose-ringed parakeet (native range)

The male rose-ringed parakeet (above, right) is distinguishable from the female (above, left) by the rose and blue colors on its head and neck and by the narrow black band running from neck to nape.

They are less than 12 inches (30 cm) in length, half of which is accounted for by the tail, and are mainly green with blue in the wings and blue sometimes in the tail. They usually move about in parties of six to eight but come together in large flocks as particular seeds ripen. There are related species in New Zealand and some of the islands of the southwestern Pacific. Ground parakeets are found in the southwestern Pacific, New Caledonia and the Loyalty Islands. These birds are up to 14 inches (35 cm) long and most species have a crest. The New Caledonia crested parakeet, *Enicognathus cornutus*, has a crest of two black feathers tipped with red, but in other species the crest is made up of six feathers, which usually are green.

The ground parakeets of Australia are of two kinds: long tailed and short tailed. The long-tailed ground parakeet, *Pezoporus wallicus*, is sometimes called the ground parrot or swamp parrot because it tends to roost in trees in swamps. It is about 12 inches (30 cm) long, half of which is accounted for by the tail. The primary color is green, with red on the forehead, and the body plumage is mottled with bands of black and yellow. The short-tailed parakeet or night parrot, *Geopsittacus occidentalis*, is nocturnal, emerging at sunset to feed. This species is now very rare, and there are fears it may be on the way to extinction, although 100 years ago it was widespread in inland Australia.

Parakeets usually roost in trees, and hanging parakeets sleep head downward, in the manner of bats. Scientists believe that many parakeet species sleep in this way. One such species is the lineolated parakeet, *Bolborhynchus lineola*, which feeds at twilight. By day it is quiescent, freezing for long periods at a time. At other times it may perch lengthwise along a branch, as nightjars do.

Some parakeets move in parties of about half a dozen, while others form flocks of hundreds, occasionally thousands. They feed in trees or on the ground, the proportion of time given to each varying according to the species. Their food is mainly seeds, fruits, leaves and flowers, and where they are numerous they are likely to become a menace to crops.

Devoted families

Parakeets usually nest in hollows in trees or among rocks, with no more than rotten wood litter on which to lay their eggs. The gray-breasted parakeet is unusual in that it builds a nest in the branches of a tree. Several may nest close together, each nest made of a mass of sticks, roofed in and with an entrance to one side, high up in a tall tree. Both female and male build the nest, but the hen alone incubates the eggs, a characteristic feature of parakeets, and the male attends her and guards the nest. The number of eggs varies from 4 to 10, according to the species. Incubation usually lasts a month or more but may be much less in some species. The chicks stay in the nest for 4–5 weeks, sometimes up to 10 weeks, and are fed by both parents.

Extinction of a species

Images of the Carolina parakeet, *Conuropsis carolinensis*, have been found in prehistoric art and at one time the bird was well established in North America, ranging from South Virginia to Texas and southeast to Florida. The Carolina parakeet had a green, rose, yellow and white plumage, but the raucous calls that it gave in flight attracted attention as much as its color. It proved easy to tame and became widely in demand as a cage bird. Moreover, the bird's flesh was considered a delicacy, while its feathers became popular adornments for women's hats. However, the Carolina parakeet was also destructive to fruit and cereal crops and consequently became regarded as an agricultural pest. If one was shot, its companions flew off but returned again and again to the spot, as if trying to rescue it, a fact that made it easy for farmers to kill large flocks. It was also killed for sport.

By the 1880s the Carolina parakeet was becoming rare in the wild, due to a combination of the factors listed above as well as the deforestation of its natural habitat for agricultural use. Ornithologists now accept that the specimen that died in Cincinnati Zoo in 1918 was the last of its kind. The Carolina parakeet was the only parakeet native to eastern North America.

The elegant grass parakeet, **Neophema elegans,** *is found in pairs or small groups during its breeding season. During the winter it often forms flocks of more than 100 birds.*

PARAMECIUM

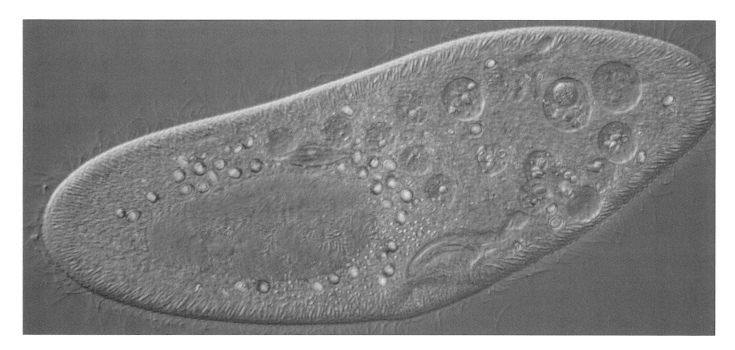

NEARLY 6,000 SPECIES OF CILIATE proto-zoans have been described. They are single-celled animals that typically, although not always, feed and swim by means of cilia. One of the most familiar types to science is the paramecium, described by Christian Huygens in 1678, when microscopy was in its infancy.

There are at least 11 species of paramecia differing in size, shape and other details, but all are elongated and more or less rounded at the ends. The largest species are about 0.3 milli-meters long and the smallest 0.05 millimeters. Each paramecium has a wide, shallow groove on its side, leading into a blind gullet. The parame-cium does not change shape like an amoeba, except when squeezing through holes, and it swims with its short cilia arranged in longitu-dinal rows. These cilia are like those that line our own breathing tubes and are associated with an orderly arrangement of structures just below the surface. The covering, or pellicle, is uniform except in the gullet and is made up of a lattice of polygons (usually hexagons) in dense rows just under the outer membrane. From the center of each emerges one cilium, or two in *Paramecium bursaria*, and at the base of each cilium is a cylin-drical basal body. To the left of each row of basal bodies, and connected to them, is a fine fiber that may be involved in coordinating the beating of the cilia. Many of the details are only revealed by electron microscopy. Among the cilia, just under the surface, are little carrot-shaped bodies called trichocysts. Each one can discharge a rod 0.02–0.04 millimeters long and ending in a sharp tip.

The trichocysts appear to help anchor paramecia when feeding, but since they are discharged when the animal is prodded, it may also be that they have a defensive function.

Within the body are two kinds of nucleus: a large macronucleus and one or more small micronuclei. For bailing out excess water, there are contractile vacuoles. In *P. caudatum* there is one at each end and the two contract alternately, but in other species there may be up to seven. Around each contractile vacuole are several radial canals that gradually fill with fluid before discharging into the central vacuole, which collapses as it discharges its contents.

Trial-and-error navigation

The cilia of paramecia are coordinated in swim-ming, each beating a little in advance of the one behind it in the line, so that waves seem to pass from the front to the rear. As the body moves through the water, it rotates counterclockwise, as seen from behind. When it meets an obstruction or an unfavorable chemical, the cilia beat in the opposite direction so the animal backs up, then turns and advances again, repeating this until a clear path is found. It has been shown that the reverse beating when the fore end is touched is brought about by a temporary increase in the permeability of the surface to calcium ions, and that stimulation of the rear end causes faster swimming by an increase in the permeability of the surface to potassium ions. The two reactions are accompanied by changes in the voltage across the surface of the animal.

A view of Paramecium caudatum, *magnified 160 times, clearly showing the hairlike cilia that help the animal to move through the water.*

Although some species occur in brackish water, paramecia live mainly in fresh water, particularly where there is abundant decaying organic matter and its attendant bacteria. They are easily cultured in infusions of hay.

Sometimes a paramecium loses much of the organization of its body and coats itself with a thick, angular membrane. This formation of a cyst may aid dispersion and survival during a period of drought, but it has not often been seen.

Temporary feeding organs

Paramecia eat bacteria, tiny algae and smaller protozoans wafted into the gullet by the cilia of the oral groove. The food is propelled down the gullet by more cilia, some of which are fused in rows to form undulating membranes. At the bottom of the gullet the food gathers in vacuoles in the cytoplasm. These tiny cavities then move into the cytoplasm and follow a path around the inside, finally discharging indigestible matter on the surface. If a paramecium is fed on dried milk stained with a dye, the dye can be seen changing color as the vacuole moves around, showing that it is being subjected first to acid conditions and then to alkaline conditions. One species, *P. bursaria*, is colored green by the cells of an alga, *Chlorella*, living within it. Both organisms are assumed to benefit from the association.

Sex at its simplest

Paramecia usually reproduce by splitting in two, a process known as binary fission. The oral groove disappears, the nuclei divide, new contractile vacuoles form and a furrow appears across the body. This process may occur one to five times a day. In addition, there occurs less frequently and often shortly after a depletion of food a form of sexual reproduction involving conjugation. In *P. caudatum*, which has one of each kind of nucleus, the two individuals join in

the regions of their oral grooves. The macronuclei break down, and in each individual the micronucleus divides into four. Of these, three degenerate while the other divides again, so each individual has two micronuclei, each with half the usual number of chromosomes. One of these migrates into the other animal to fuse with its opposite number. The two animals then separate and the single effective nucleus in each divides again three times to give eight nuclei from which are formed the micro- and macronuclei of four new individuals. Sometimes self-fertilization occurs: the two animals unite but do not exchange nuclei. In other instances self-fertilization occurs without mating. There are no males and females, but each species occurs in a number of varieties and each variety in more than one mating type; conjugation occurs only between members of different mating types.

PARAMECIA

PHYLUM	**Ciliophora**
CLASS	**Nassophorea**
ORDER	**Nassophoria**
SUBORDER	**Peniculina**
FAMILY	**Parameciidae**
GENUS	***Paramecium***

LENGTH
0.05–0.3 mm

DISTINCTIVE FEATURES
Large, complex single-celled organism; covered in hairlike cilia used in movement and feeding; groove on side of body leading to primitive mouth, gut and anus

DIET
Smaller microorganisms such as bacteria, algae and tiny protozoans

BREEDING
Asexual reproduction by binary fission, or sexual reproduction by conjugation followed by binary fission

LIFE SPAN
Varies according to species and environment

HABITAT
Fresh water; some species in brackish water

DISTRIBUTION
Worldwide

STATUS
Superabundant

Two individuals of Paramecium caudatum in the process of conjugation, a sexual encounter in which the micronuclei divide, mutate and migrate from one body to the other.

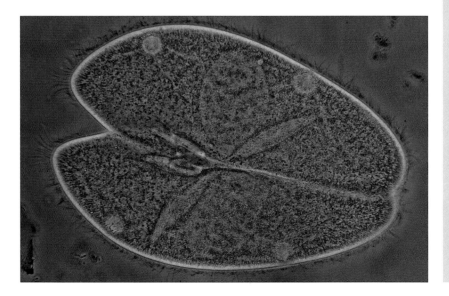

PARROT

ONLY ABOUT ONE-THIRD of the 332 members of the parrot family are referred to by the name parrot. The rest are known by a variety of other terms, including budgerigar, cockatoo, kea, lorikeet, lory, lovebird, macaw and parakeet (all are described elsewhere in this encyclopedia). The parrots include amazons, stout-bodied American birds with mainly green plumage and short, square or rounded tails. One of the largest is the yellow-headed amazon, *Amazona ochrocephala*, which is 15 inches (37.5 cm) long and is green except for the head and some blue and red in the wings. It ranges from Mexico to Brazil. One of the smallest amazons is the white-fronted amazon, *A. albifrons*, which is 10 inches (25 cm) long and has a white forehead and bright red lores; the male has a red wing patch. The African gray parrot, *Psittacus erithacus*, which lives in the forests of tropical West and Central Africa, has white cheeks and a red tail.

All members of the parrot family have large heads, short necks and hooked bills, with the upper mandible longer than the lower mandible and curving downward. There is a broad cere (a waxy protuberance at the base of the bill) through which the nostrils open. Of the four toes on each foot, two are directed forward and two backward. Powder down feathers are scattered through the plumage.

Raucous and agile

Parrots are essentially forest dwellers, traveling in noisy flocks, the smaller parrots twittering and the larger birds uttering raucous shrieks and squawks. They climb around the trees, using their bills as well as their feet, and, unlike most other birds, can hold food with one foot. Parrots feed on fruits, seeds and nuts. They seem to have a more highly developed sense of taste than most birds and test food with their fleshy tongues before eating it.

The amazons and the African gray have been favorite pets for centuries. This is primarily because of their being able to talk, and the main interest in parrots lies in their vocal mimicry. A good talker can imitate almost any sound, from the songs of other birds and mechanical sounds

to human speech. It may be able to whistle tunes, sing short phrases from songs, mimic laughter or crying and even call people by name: in short, express itself vocally in a remarkably humanlike way. An African gray has been recorded as imitating actions, using the foot to imitate its owner's use of the hand. Some African gray parrots have been reported to have acquired a vocabulary of more than 700 words, the equivalent of that of a seven-year-old child.

Avian vocal dexterity

The reputation of parrots as talkers has suffered in recent years from claims that budgerigars, as well as mynahs, have greater abilities in this respect. However, it is doubtful whether any of the three birds is more skilled at mimicry than the others. In the past there was considerable debate about the extent to which a parrot was

Eclectus parrots, Eclectus roratus (male shown above), are sexually dimorphic birds. The female is bright red and blue in color, with a black bill. In most other parrots the sexes look alike.

ORANGE-WINGED AMAZON

CLASS	**Aves**
ORDER	**Psittaciformes**
FAMILY	**Psittacidae**
GENUS AND SPECIES	***Amazona amazonicus***

WEIGHT
10½–16½ oz. (300–470 g)

LENGTH
Head to tail: up to 12½ in. (31 cm)

DISTINCTIVE FEATURES
Stout body; large, thickset head; strong, hooked bill; short, square tail; yellow forecrown and area from base of bill to behind eye; blue stripe over and behind eye; remainder of head blue green; body mainly green; orange and yellow patches on wings; tail tipped with yellow

DIET
Fruits such as oranges, mangoes and cocoa

BREEDING
Age at first breeding: probably 1 year; breeding season: December–February (Colombia); number of eggs: 2 to 5; incubation period: about 21 days; fledging period: about 60 days; breeding interval: probably every 2–3 years

LIFE SPAN
Up to about 40 years

HABITAT
Various habitats below 2,000 ft. (600 m), including deciduous forest, secondary growth, savanna with tall trees, parkland and gardens; roosts in giant bamboo

DISTRIBUTION
Amazon Basin north through Venezuela to eastern Colombia, west to eastern Ecuador and Peru

STATUS
Abundant: commonest large parrot in much of its range

Amazons are stout parrots with mainly green plumage, found in the tropical forests of Central and South America. Reaching a maximum length of 12½ inches (31 cm), the orange-winged amazon is one of the smallest species.

aware of the meaning of the humanlike sounds it could produce. In fact, careful study shows that parrots associate words and sounds with events and persons in much the same way, and to much the same degree, as a child of two years of age. Parrots are also believed to imitate sounds in the wild. A tame Eurasian jay, *Garrulus glandarius*, is also a good talker, and wild jays have been recorded as imitating the hoot of an owl.

The structure of a bird's voice box, or syrinx, makes the fact that it can imitate such a wide range of sounds quite remarkable. The human larynx, with its vocal cords, is near the top of the windpipe. Air passing across the vocal cords makes them vibrate. The sounds produced are modified to form words and other modulated sounds by alterations in the position of the tongue and the teeth and by changes in the shape of the cheeks and lips. A bird's voice box is at the base of the windpipe and is worked by a dozen small muscles to produce modifications in sound. There are no cheeks or lips, in the sense of a human's pliable mouthparts, there are no teeth and the tongue is less mobile than that of a human. As a result, a bird must produce all of its sounds far down in the throat, which explains how it is able to sing with a closed bill. This also explains how a parrot is able to talk with its bill nearly closed.

Orange-winged amazon

Cavity nesters

The breeding habits of parrots are fairly uniform. The birds usually nest in cavities in trees, but some use burrows in the ground or crevices in rocks. A few Australian parrots dig holes in termite mounds. The ground parrot cuts thin stems of sedges and rushes to form a nest cavity at the center of a clump of vegetation. It hollows a scrape in the sand, lines it with cut herbage and makes one or two access tunnels through the vegetation. The nesting cavity is not lined; at best, the eggs are laid on the powder of rotten wood or similar material. The eggs are nearly spherical, white and somewhat glossy. The clutch varies from 2 to 3 in the larger species to 11 in the smaller species, such as the green-rumped parrotlet, *Forpus passerinus*.

The female African gray lays her eggs at intervals of 3 days, each clutch usually consisting of four eggs. Incubation, which is carried out by the hen alone, lasts just over a month. The newly hatched chick is flesh-colored with a pale bill and pale claws. The body soon becomes covered with a light down and the bill turns black after a few days. The hen broods the chicks, especially at night, for 2 months, the male sharing the feeding of the nestlings by regurgitation. When first fledged, the young parrot resembles the parent except that the body feathers are a darker and softer gray, the tail feathers are a less vivid red and the eye is black. The female orange-winged amazon, *A. amazonica*, lays two to five eggs in a hole in a tree. She incubates them for about 3 weeks and attends the chicks for about 2 months.

A globally threatened family

Parrots have been popular pets for millennia due to their colorful appearance and mimicry abilities. They were kept in ancient Greece and Rome and explorations of the New World during the 15th and 16th centuries, along with subsequent discoveries in Southeast Asia and Australia, made many more species known. Recently the trade in parrots both as pets and as collectors' items, along with the destruction of their native rain forest habitat, has resulted in a decline in the populations of many species. The BirdLife International checklist of globally threatened birds lists 86 species of parrots as threatened. It is almost certain that two of these species, as well as at least nine more species and eight sub-species, have become extinct since 1600.

Parrots, such as these orange-winged amazons, regularly visit earth banks to lick the clay, which seems to help absorb and filter out harmful toxins in their diet.

PARROTBILL

Balancing precariously among the slender reed stems, a female bearded reedling searches for insects and other sources of food, such as the seeds of reeds, which form its staple diet in the winter months.

THESE SMALL TITLIKE BIRDS have a short, deep convex bill, like that of a miniature version of a parrot. They are also known as crow tits. Some ornithologists class parrotbills with the tits; others group them with the babblers. Most of the 20 species live in eastern Asia, from India to China but not in the Malayan region. However, the bearded reedling, *Panurus biarmicus*, ranges from China to Britain, where it breeds in a few areas mainly on the east coast. This species was formerly called the bearded tit, the male having a prominent black stripe on each side of the bill like a drooping mustache.

The bearded reedling is about 5 inches (12.5 cm) long with a 2¾-inch (7-cm) tail and looks like a small, long-tailed sparrow. The plumage is tawny brown with black, white and brown stripes on the wings. The bill is yellow and the head is blue gray. The female's plumage is dull by comparison.

A little larger than the bearded reedling is Gould's parrotbill, *Paradoxornis flavirostris*, also known as the black-breasted parrotbill, which is 7 inches (18 cm) long, with brown plumage above and whitish plumage below, and black patches around the eyes. Blyth's parrotbill, *P. poliotis*, is only 4 inches (10 cm) long.

Agile acrobats

Parrotbills clamber agilely among grasses and bamboo in a very titlike manner, pausing straddled between two stems. The habits of many of them are hardly known, but most live in groups of up to 50, sometimes with other birds such as tits and babblers. They forage in tall grasses, bamboo or the lower levels of trees, where their constant chattering gives them away.

The bearded reedling lives in dense reed beds, its presence indicated by its high-pitched calls. Often all that will be seen is a glimpse of a small dark bird disappearing into the reeds. However, particularly in the fall, small groups can be seen flying over the reeds or foraging or roosting among them.

Stripping reeds

Parrotbills feed on insects, grass seeds and sometimes berries. During the breeding season reedlings feed on insects, caterpillars, mayflies and other insects, together with a few small snails. In the winter, however, they feed mainly on the seeds of reeds. The black-throated parrotbill of Nepal, Myanmar (Burma), Thailand and China feeds on bamboo shoots as well as insects. The reed bed parrotbill, *Paradoxornis heudei*, feeds on insects and their larvae that live inside the stems of reeds. It is restricted to extreme southeastern Siberia and reed beds in eastern China. It searches for the holes made in the reeds by insects and then straddles the reeds, inserting the tip of the upper mandible and tearing away strips of reed until the insects are exposed.

Reed hammocks

Parrotbill nests are made of the leaves of grass or bamboo or strips torn off reeds woven tightly into cups around bamboo or stout grass stems and bound with cobwebs. Reedlings line their nests with reed flowers and a few feathers. Both sexes help in building the nest, but the male makes the lining. When he is displaying to the female, he fluffs out his mustaches and erects a sort of crest on his head. The pair posture to each other with their tails spread, and fly up together with quivering wings. In captivity the female reedling has been seen roosting under the male's wing. Reedlings do not hold territories but search for food over a large area of reed bed.

BEARDED REEDLING

CLASS	**Aves**
ORDER	**Passeriformes**
FAMILY	**Paradoxornithidae**
GENUS AND SPECIES	***Panurus biarmicus***

ALTERNATIVE NAME
Bearded tit

LENGTH
Head to tail: about 5 in. (12.5 cm)

DISTINCTIVE FEATURES
Short, yellow bill like that of a miniature parrot; very long tail. Male: blue-gray head; black mustache; tawny brown plumage. Female: duller overall, with brown head.

DIET
Summer: mainly invertebrates; late fall and winter: mainly seeds; mixture at other times

BREEDING
Age at first breeding: 1 year; breeding season: late March–July; number of eggs: 4 to 8; incubation period: 11–12 days; fledging period: 12–13 days; breeding interval: 2 to 4 broods per year

LIFE SPAN
Up to 6 years

HABITAT
Reed beds close to fresh or brackish water

DISTRIBUTION
Scattered range in Europe; Caspian Sea east to northeastern China

STATUS
Locally common within specialized habitat

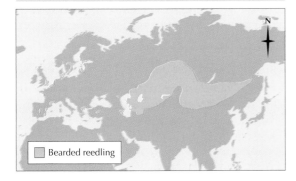

Bearded reedling

The eggs, two to four for most parrotbills but four to eight for bearded reedlings, are incubated for nearly 2 weeks (11–12 days in the case of bearded reedlings) by both parents. The chicks stay in the nest for about 10–13 days. There are two, sometimes three or four, broods a year.

Resilient reedling

The bearded reedling was for many years one of the rarest British breeding birds. At one time it was much more widespread in southern and eastern England. However, cutting of reed beds, drainage of marshes and the sale of both the eggs and the birds cut the range of the reedling until it bred only in the reed beds of Norfolk and Suffolk and in a few other scattered localities. Since the 19th century the bearded reedling has been safe from human persecution, but it has suffered badly during hard winters. After the long winter of 1916–1917 very few bearded reedlings were left. Numbers increased again until 1939–1940, when there was another severe winter. Again they recovered. However, so few reedlings survived the hard winter of 1947 that it was thought they would become extinct in Britain. In the summer of 1947 only four breeding pairs were recorded. Yet once again there was a remarkable recovery, and 10 years later over 100 pairs were recorded in the counties of Norfolk and Suffolk. Since then numbers have continued to increase, with only temporary setbacks caused by bad weather or flooding, and reedlings have now spread to other parts of England. Groups of reedlings can be seen leaving the reed beds after the breeding season. They spread across the land in search of new reed beds where they can feed, and some stay there to breed. Bearded reedlings have also crossed the North Sea from Holland.

The severe effects of hard winters on the reedling population is due to snow covering the reeds. In 1962–1963 the population was not affected as much as in 1947 because, although the frost was harder and lasted longer, there was less snow and the reedlings were able to find food. There are now more than 600 pairs in Britain, which is a dramatic change in their fortunes.

With its prominent drooping mustache, it is easy to see how the bearded reedling (male shown above) got its name. Bearded reedlings such as this range from China to Britain, but within this large range they are highly localized.

PARROTFISH

The bullethead parrotfish, **Scarus sordidus**, *is named for its blunt, squared-off face. The mouthparts of the male (above) are greenish, while those of the female are pink.*

PARROTFISH ARE SO CALLED because of their teeth, which are joined to form a "parrot's bill" in the front of the mouth. There are 83 species of parrotfish, grouped into 9 different genera. They vary in adult size from 1–6 feet (0.3–1.8 m), although a few individuals have been measured at 12 feet (3.7 m) long. At cruising speed, parrotfish swim with their pectoral fins, using their tail only to swim more quickly.

Parrotfish of the genus *Sparisoma* may live solitary lives or may come together in small groups without any social organization. Some species of the genus *Scarus* move about in large schools of up to 40 when feeding, rather like herds of cattle. The schools are made up of fish of about the same size. In some species of *Scarus* the groups are smaller and are made up of several females with a mature male acting as leader, in the same way that cows do with a bull. If another male joins the group, he is chased away, the alpha (dominant) male sometimes trailing him at a distance of 20 feet (6 m) for perhaps 100 yards (91 m) before rejoining his harem.

Sexual coloration

For a long time parrotfish were separated into species on the basis of color. Then zoologists discovered that the same species could appear in different colors. Some species go through at least three different color phases in the course of a lifetime. In others there is a marked difference between male and female. Zoologists originally regarded the princess parrotfish, *Scarus taeniopterus*, which bears orange and blue stripes, as a separate species from the striped parrotfish, *S. croicensis*, also referred to as *S. iserti*, which is striped brown and white. Subsequently they realized that both belong to the same species, the first being the male and the second the female.

When males and females of a species are differently colored, the young fish are colored like the females. The young females keep these colors as they mature, but the males take on the colors of the mature males. In some parrotfish, the males become bumpheaded over time. Instead of the forehead sloping, it becomes a large bump. Old males have heavy, blunt snouts.

Homing by the sun

Some parrotfish spend the nights under overhanging ledges of rock or in caves. If they are alarmed during the day, they swim straight for their night quarters. To try to find out how the fish were able to home in so accurately on their destination, scientists carried out a series of experiments. First, they suspended a net in front

PARROTFISH

CLASS **Osteichthyes**

ORDER **Perciformes**

FAMILY **Scaridae**

GENUS **9 genera**

SPECIES **83 species, including bumphead parrotfish, *Scarus perrico* (detailed below); bullethead parrotfish, *S. sordidus*; princess parrotfish, *S. taeniopoterus*; and striped parrotfish, *S. croicensis* (also *S. iserti*)**

LENGTH
Up to 31½ in. (80 cm)

DISTINCTIVE FEATURES
Series of dark lines radiating from eyes; light bluish green body; darker blue fins; blue tooth plates, forming a powerful bill; large individuals of both sexes have massive fleshy hump on head over eyes

DIET
Mainly living corals; also algae, particularly coralline red algae

BREEDING
Young females significantly outnumber young males at hatching; males mature into primary (small) or secondary (large) adults; females undergo sex change into males if too few secondary males are present

LIFE SPAN
Not known

HABITAT
Coral reefs at depths of 10–100 ft. (3–30 m)

DISTRIBUTION
Eastern Pacific: central Gulf of California and northern Mexico south to Peru, including Galapagos Islands

STATUS
Common

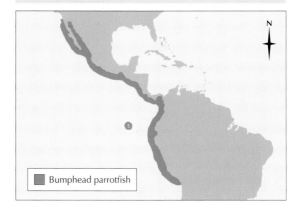

Bumphead parrotfish

of a cave. When the parrotfish were disturbed, they swam straight for the net and continued trying to swim through it. After the net was lifted, they swam straight into the cave. In the next test, the scientists netted some parrotfish that were known always to swim in a southeasterly direction to their caves. They were then taken farther along the coast and put back into the sea. They immediately swam on a southeasterly course, despite the fact that there were no caves in this direction. When this experiment was repeated, scientists noted that if a cloud passed across the sun, the fish became temporarily confused. They swam about in different directions until the sun shone again, when they once more swam unerringly on a southeasterly course. Finally, the scientists blindfolded the fish by putting suction caps over their eyes. The fish subsequently swam in all directions in a disoriented state, but when the caps were removed, they swam straight for their caves. From these tests, the scientists concluded that parrotfish need the sun in order to navigate.

Herbivorous diet

With their bill-like teeth parrotfish browse eelgrass and seaweed, often nipping pieces off the tropical coral reef that forms their native habitat. Most species also have a second set of flattened pharyngeal (throat) teeth that enables them to grind down coral reef. The undigested coral fragments are passed out and dropped in special places along the route the parrotfish follow to and from their caves, accumulating in heaps. Some parrotfish, in particular the large parrotfish, *Chlorurus microrhinos*, excavate large amounts of coral, and bring about significant erosion of reefs. The bumphead parrotfish, *Scarus perrico*, eats large amounts of live coral.

By feeding on coral reefs, species such as the redlip parrotfish, Scarus rubroviolaceus, contribute significantly to natural reef erosion.

The teeth of parrotfish are joined to form a bill in front of the mouth. Female parrotfish, such as the rusty parrotfish, Scarus ferrugineus (above), are generally drab red, brown or olive green in color.

Several kinds of reproduction

The behavioral patterns of parrotfish, including their breeding habits, vary substantially from one species to another. The eggs of *Scarus* species are elongated and oval, whereas those of *Sparisoma* are spherical. The redfin parrotfish, *S. rubripinne*, the spawning habits of which have been closely studied by scientists, was found to breed year-round but only in the afternoons. At this time a milling mass of fish leaves its feeding ground close inshore and assembles in depths of 65–70 feet (9.8–21.3 m), the mass keeping a few feet up from the bottom. Most of the fish assembled there are males, and while the spawning is going on there is a preponderance of females inshore. Every so often groups of 4 to 13 fish swim upward from the main mass and circle around to release eggs and milt (sperm-containing fluid).

The bulk of the young result from this method of reproduction, although there is a second type of spawning in which a solitary male mates with a solitary female. In another species a male and a female swim up to the surface, circling around each other as they go. As they get near the surface, they rotate around each other, then release a cloud of eggs and milt. The eggs of all parrotfish species contain an oil drop, enabling them to float near the surface. They range from 1–25 millimeters in diameter, and hatch in a day, releasing fish larvae. At hatching, there are more females than males and both sexes bear identical color patterns.

Male parrotfish that mature into small adults are known as primary males; those that mature into larger individuals are referred to as secondary. The secondary fish tend to be solitary and maintain a sexual territory with a certain number of females. In communities where not enough primary males become secondary, some females, usually the larger individuals, undergo a sex change into a secondary male. This process takes a few days and is not reversible.

Nocturnal cocoon

Some species of parrotfish, including the rainbow parrotfish, *Scarus guacamaia*, give out mucus, or slime, from glands in their skin as night falls. The mucus forms a kind of loose, foul-smelling covering, with an opening in the front guarded by a flap that allows water in and a hole at the back that lets it out. The parrotfish is thus able to draw water in and pass it across its gills to breathe even while it is enclosed. In the morning the parrotfish breaks out of this mucus cocoon and goes about its normal activities. When a parrotfish rests at night, its breathing drops to a low rate and it enters a sleeplike state.

The mucus envelope may help to prevent the gills from silting up while the fish is resting on a sandy bottom. Alternatively, it may provide the parrotfish with protection from predators. Its unpleasant smell may well deter nocturnal predators that hunt by scenting their prey, thus protecting the sleeping parrotfish.

PARTRIDGE

THE GRAY PARTRIDGE IS famous as a game bird on cultivated farmland in Europe and in Central Asia east to the Caspian region. There are, however, many related species in Europe, Asia and Africa, including tree partridges, wood partridges and bamboo partridges.

The gray partridge, *Perdix perdix*, known as the Hungarian partridge in North America, is a plump bird with a rounded body about 1 foot (30 cm) long. It has a short bill and a short tail and weighs ¾–1 pound (340–450 g). Its plumage is generally brownish with gray and white markings. The back of the male is brown streaked with buff and white. The neck is gray and the head is orange. However, the gray partridge's coloration is quite variable and it is probably best distinguished by the chestnut colored horseshoe on its pale gray lower breast. The head of the female is more streaked than in the male and the horseshoe is smaller, or absent.

The red-legged partridge, *Alectoris rufa*, is slightly larger, 1⅛ feet (34 cm) long, and can be distinguished from the gray partridge at close quarters by its white cheeks and throat, bordered by a black band. Its flanks are barred with black, white and chestnut, and its bill and legs are red compared to the bluish gray of the gray partridge.

The gray partridge, with its several subspecies, ranges across most of Europe into western Asia. It has also been introduced into the United States and Canada. The red-legged partridge is native to southwestern Europe, but has been introduced into several more northerly countries in Europe.

Other species

There are 18 species of tree partridges, genus *Arborophila*, found living in forests from eastern India to southern China and Peninsular Malaysia. All are attractively marked, as are the five species of wood partridges, genus *Rollulus*, of Peninsular Malaysia to Borneo. The three species of bamboo partridges, genus *Bambusicola*, of southern China to Myanmar (Burma) and Vietnam, have long tails and spurs and look more like pheasants.

Partridges are usually found in small parties, known as coveys. The photograph above shows a covey of rock partridges, Alectoris graeca. *In this genus both sexes have red legs and a red bill.*

A pair of adult gray partridges dust bathing. This behavior removes parasites and helps to clean the feathers.

GRAY PARTRIDGE

CLASS	**Aves**
ORDER	**Galliformes**
FAMILY	**Phasianidae**
GENUS AND SPECIES	***Perdix perdix***

ALTERNATIVE NAME
Hungarian partridge (North America only)

WEIGHT
¾–1 lb. (340–450 g)

LENGTH
**Head to tail: 11½–12¼ in. (29–31 cm);
wingspan: 17¾–19 in. (45–48 cm)**

DISTINCTIVE FEATURES
Plump body; appears neckless. Male: orange face; brown-gray underparts with chestnut flank barring; chestnut horseshoe on belly; reddish tail feathers. Female: more streaks on head; horseshoe smaller or absent.

DIET
Grass and cereal leaves; grain and wheat seeds; roots and shoots; occasionally insects

BREEDING
Age at first breeding: 1 year; breeding season: April–September; number of eggs: 10 to 20; incubation period: 23–25 days; fledging period: about 15 days; breeding interval: 1 year

LIFE SPAN
Up to 5 years

HABITAT
Grassland not much taller than bird's head, with some taller and denser cover nearby such as hedgerows, woodland verges or scrub

DISTRIBUTION
Much of Europe and Central Asia; introduced to parts of North America

STATUS
Common, but declining in many areas

Party life

The gray partridge usually walks with its neck drawn into its shoulders, giving it an almost neckless appearance. However, when suspicious it stretches its neck up to keep watch and then runs for cover. It lives in small parties, or coveys, except during the breeding season. The partridge's preferred habitat is grassland not much taller than the bird's head, with some taller and denser cover nearby such as hedgerows, woodland verges or scrub. When suddenly startled, the whole covey rises with a whirring of wings and flies strongly and swiftly, although not for long distances, typically hedgehopping.

The red-legged partridge seems to be more restless in its habits and moves more quickly than the gray partridge. It also tends to run rather than fly when alarmed, and the covey scatters instead of keeping close together as gray partridge coveys do.

Partridge coveys are made up of adult birds and their young, and these stay together until the next breeding season. Several separate coveys of gray partridges sometimes join together to roost among the ground vegetation.

The chief call of a gray partridge is a hoarse note, rather like a key being turned in a rusty lock. That of the red-legged partridge is a *chack-chack-chack*, like a steam engine.

Diet of grasses and seeds

Partridges are fairly inactive most of the day, although they may feed a little. Their main feeding periods are early morning and in the evening. Their food is almost entirely made up of plant material such as grass and cereal leaves,

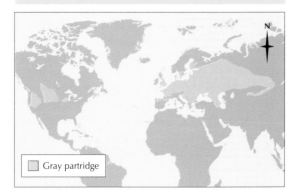

Gray partridge

grain and weed seeds, young shoots, some roots and fallen berries, and sometimes root crops such as turnips and beets. Very occasionally they will eat insects, mainly ants and their pupae.

Pair for life

Gray partridges form monogamous pair-bonds and where both survive, the pair commonly reunite the following season. They pair up to 4 months before the breeding season, which is from April to September. The nest, a scrape in the ground lined with dry grass and dead leaves, is built among tall herbage or under bushes or other cover. The female lays her eggs, usually 10 to 20 in a clutch, in April or May at intervals of 24–36 hours. The somewhat pear-shaped eggs of the gray partridge are olive brown, while those of the red-legged partridge are yellowish white to pale yellowish brown in color, with a few reddish spots. The gray partridge covers her eggs with grass until the clutch is complete, but the red-legged partridge leaves hers uncovered. Incubation by the female begins when the clutch is complete, and all eggs hatch within a few hours of each other after a period of 23–25 days.

Both parents tend the young, which are also initially brooded by both parents. The male often broods the first chicks to hatch while the female continues to incubate the remaining eggs. The chicks begin to flutter when 10 days old and fly at about 15 days. However, the association between the parents and their young continues into the fall and winter. The oldest known gray partridge studied lived for 5 years.

Numerous predators

Little is known about predators, but partridge eggs and chicks are probably taken by ground predators such as foxes and ermines, as well as by birds such as crows, rooks, magpies and buzzards. Any bird that lays up to 20 eggs or more in a clutch can be assumed to have numerous predators or the species would soon overrun its habitat. The gray partridge is still common across most of its range, but is now declining in many areas of intensive agriculture.

Gray partridge chicks fledge at a very young age, but remain in coveys with their parents until the next breeding season.

PASTORAL ANT

Honey farms

Some ants take honeydew only when they happen to come across aphids. In others milking is less haphazard, and the ant actually strokes an aphid with its antennae, stimulating the aphid to increase the flow of honeydew. Ants also establish aphid farms by shepherding the aphids to the more tender parts of a plant. The move to this better pasture results in the aphids giving a greater supply of honeydew, so benefiting both the ants and the aphids.

The link between the two insects is so close that if the aphids are threatened, the ants will pick them up and carry them away to safety. It is thought that some ants even take aphids into their underground nests for the night. They bring them out again the following morning and place them back on the leaves from which they were taken the previous evening.

Providing shelter

It is not only green leaves that serve as pasture. There are ants that take aphids underground, excavating cavities for them, to feed on the sap in roots. Other ants build special shelters on the stems of plants that may be just roofs or may be complete stables, entirely enclosing the aphids. The ants use various materials for such building. For example, earth is mixed with saliva and made into a cement and chewed plant fibers make a form of paper, known as carton. The ants manufacture this paper in much the same way that wasps make their paper nests. The making of carton is a feature more especially of certain tree ants, genus *Crematogaster*, which also use it for making brood chambers in their nests for their own eggs and larvae.

Egg kidnapping

Some ants carry their pastoral activities a stage further. In the fall aphids lay their eggs on the stems of shrubs. Normally the eggs remain there all winter, hatching the following spring. However, not all the eggs survive because of their many predators, including some of the

Pastoral ants, such as this black ant, Acanthomyops nigra (above, center), not only milk aphids for honeydew, but often herd them to the most nourishing parts of plants so that the aphids will provide a greater supply of honeydew.

PASTORAL ANTS DO NOT represent a natural classification but are, rather, an ecological grouping of those species of ants that exhibit a form of husbandry. It has long been known that some species of ants regularly visit aphids, otherwise known as greenflies or blackflies, to take honeydew from them. This habit is so well known today that there are probably few people who have not heard it. What is not so well known is that not only do these ants milk the aphids, much as humans milk cows, but they also tend, herd and shepherd them, much as livestock are cared for by human beings. Moreover, other insects such as scale insects, leafhoppers and the caterpillars of many butterflies are also used by ants in this way because they give out either honeydew or some other sweet fluid.

PASTORAL ANTS

PHYLUM **Arthropoda**

CLASS **Insecta**

ORDER **Hymenoptera**

GENUS AND SPECIES **Many; pastotal ants are a very varied ecological grouping rather than a strict scientific classifiction**

LENGTH
Depends on species and caste

DISTINCTIVE FEATURES
Varies according to species

DIET
Honeydew from aphids or other suitable insect hosts

BREEDING
Winged females mate with winged males; each queen (fertilized female) starts a new colony, tending first batch of eggs and larvae

LIFE SPAN
Depends on species and caste

HABITAT
Anywhere that aphids or other suitable insects hosts are found

DISTRIBUTION
Almost worldwide

STATUS
Common or abundant

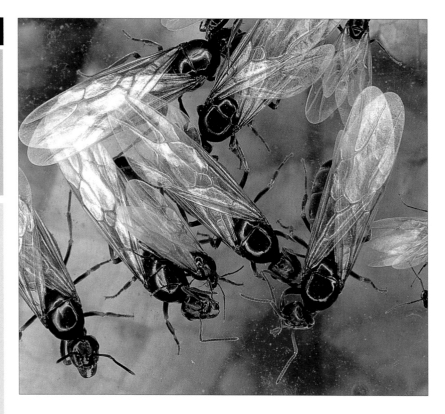

Winged queens of the black ant, Lasius niger, *a species notorious for tending aphids that feed on corn and cotton roots. Were it not for these ants, the aphids would not be such a serious pest of some commercial crops.*

small insectivorous birds that are forever searching the crevices in the bark of trees and shrubs for anything edible. During the fall there are some species of ants that wait for the aphids to lay their eggs, and then carry the eggs down into deep underground nests. The ants tend the eggs throughout the winter. Then, when the larvae hatch in the spring, they are carried up above ground and placed on plants to feed. The ants may place them first on one plant and then on another as the spring succession of green shoots unfolds, so providing the young aphids with the best possible pastures.

Life cycle

The life cycle of pastoral ants themselves is much like that in other ants. Winged females mate with winged males, after which the males die and the females shed their wings. Each female (queen) then goes on to start a new colony, laying eggs. The queen tends these first eggs and feeds the

first batch of young with her saliva. These become workers and proceed to tend subsequent eggs. Females come from fertilized eggs and males from unfertilized eggs. Whether a female develops into a worker or a queen depends the on diet that she is fed.

Overgrazing

In spring the foliage of a climbing rose, a peach tree or other shrub or tree sometimes suddenly becomes infested with aphids. One day it is to all intents clean, and then just a few days later it is covered with these plant lice. These visitations are partly the results of the high rate of reproduction of the aphids; one day they are too few to be noticeable and the next they are everywhere and the leaves are curling. In some instances, the original cause of this may be pastoral ants carrying the aphids and planting them on the bush or shrub. This is especially true of young peach trees, where it is often possible to see the ants streaming up and down the stem and along the branches, among the hordes of aphids.

One North American ant, *Lasius niger americana,* is notorious for tending the aphid *Anuraphis maidi-ridicis,* which feeds on the roots of cotton and corn. The ant tends the aphid eggs in winter and then plants the young aphids on the crop in spring. Were it not for the ants, the aphids in some places probably would not survive the winter, or would do so in such small numbers as to be negligible. As it is, the intervention of the ants can sometimes result in serious damage to crops that would otherwise be unaffected.

PATAS MONKEY

The adult male patas monkey has an almost military bearing, with his dark red coat, long, white whiskers and heavy mane.

and red body color. The male also has a mane and very long, white whiskers. In the female the light areas are fawn. Both sexes have a dark tail and a whitish face, and the nose is often blue.

Patas monkeys are found throughout the savanna zone, especially in regions of long grass and in lightly wooded areas, from West Africa to as far east as the Sudan, Uganda, Kenya and Ethiopia. Able to tolerate very dry conditions, they have extended their range into Aïr, in the southern Sahara.

Females dominate

A troop of patas monkeys is led by one of the females. The monkeys live in small troops, of between 6 and 30 animals. There is only one adult male in each troop, the males being intolerant of one another other in the presence of females. Other adult males live alone or in bachelor bands in outlying areas. When two troops meet, which is rare, the larger one generally chases away the smaller. An individual male from the bachelor band may sometimes prove stronger than the male in a troop and so take his place. When the troop male meets a strange patas or troop of them, he utters a series of deep barks. This rallies the females of the group, one of which may give chase.

The troop has a home range of around 20 square miles (52 sq km), sometimes more, in the semiarid savanna. During the day the troop wanders about over its home range, covering a distance of not less than ¼ mile (0.4 km) in a day, and sometimes nearly 8 miles (13 km). The patas monkeys feed in the morning and again in the evening, resting during the heat of the day.

Grasses an important food

Patas monkeys eat grasses, the fruit pulp of the tamarind and the fruits of several other trees, as well as seeds and berries. They pluck these with their hands, not with their mouths, and they often stand on their hind legs to pull down seeds from tall grasses or bushes. They cannot uproot grasses or dig for roots and tubers as the stronger baboons do. Patas are fond of mushrooms, especially the big *Lepiota*, and it is over these favored items that some of the few quarrels occur in patas societies. Patas monkeys also feed on some animal food, including insects and birds' eggs.

Most births in dry season

Females come into season every 30 days or so. When mating is over, the female stays with the male, and after a while they mate again. Breeding

A LARGE RED MONKEY related to the guenons, the patas monkey is also called the red guenon, hussar monkey or military monkey. It lives mainly on open grasslands and savanna. It is built a bit like a greyhound, with a deep chest, supple back and long legs enabling it to run fast. The males are on average about twice the size of females. An adult male may be up to 3 feet (90 cm) long in head and body with a 2½-foot (75-cm) tail. He weighs up to 30 pounds (13 kg), while a female will weigh only 9–15½ pounds (4–7 kg). The male is more brightly colored than the female: the rump and back of the hind legs are white, standing out against the gray

PATAS MONKEY

CLASS	**Mammalia**
ORDER	**Primates**
FAMILY	**Cercopithecidae**
GENUS AND SPECIES	***Cercopithecus patas***

ALTERNATIVE NAMES
Red guenon; hussar monkey; military monkey

WEIGHT
**Male: 15½–30 lb. (7–13 kg);
female: 9–15½ lb. (4–7 kg)**

LENGTH
**Head and body: 2–3 ft. (60–90 cm);
shoulder height: 1⅔–2 ft. (50–60 cm);
tail: 1⅔–2½ ft. (50–75 cm)**

DISTINCTIVE FEATURES
**Long legs; white thighs and face; cream or
white belly; dark tail. Male: gray and reddish
upperparts; heavy mane; very long, white
whiskers. Female: lacks mane and whiskers;
duller coloration.**

DIET
**Grasses, seeds, berries, fruits and
mushrooms; also insects and birds' eggs**

BREEDING
**Age at first breeding: 3–4 years (male),
2½ years (female); breeding season: June to
September; number of young: 1; gestation
period: 167 days; breeding interval: 1 year**

LIFE SPAN
Up to 23 years in captivity

HABITAT
**Open grassland and savanna, especially in
areas of long grass and scattered trees**

DISTRIBUTION
**West Africa east to Sudan, Uganda, Kenya
and Ethiopia**

STATUS
Locally common

Patas monkey

takes place throughout the year but there is a peak between June and September. The single young is born after a gestation period of 167 days, often in the dry season between December and February. At first the baby clings to its mother's belly, but it gradually becomes independent. Infants spend most of their time playing, chasing and wrestling and do so up to about 1 year old.

A patas monkey in captivity. These long-limbed monkeys are mainly terrestrial in habit, living in troops on open grassland and savanna.

Defensive behavior

Lions, leopards, hyenas, jackals, wild dogs, pythons, crocodiles and eagles all prey upon patas monkeys. Because he is not dangerous and fearsome like a baboon, the male patas monkey makes a diversionary display in the face of such danger. He bounces noisily on bushes and trees and then runs away through the grass, drawing attention away from the females and young. They, meanwhile, stand hidden in the long grass or flee silently. Unlike the females, the male is conspicuously colored, so he cannot hide and has to rely on his greyhound build to carry him swiftly through the grass. Patas monkeys are among the fastest in the world and can run at speeds of up to 34 miles per hour (55 km/h).

The male also acts as a watchdog. He is the first to survey an area before the troop goes in, and he often stands upright like a sentry, gazing out over the long grass and using using his tail as an extra support.

PEAFOWL

Most birds display during the breeding season, and the male Indian peafowl is no exception. Commonly known as the peacock, he uses his sweeping green train with its striking eyespots to spectacular effect during this time.

THERE ARE THREE SPECIES OF peafowl, the blue, common or Indian peafowl (*Pavo cristatus*); the green peafowl (*P. muticus*), which lives farther east, from Myanmar (Burma) to Java; and the Congo peafowl (*Afropavo congensis*). The Indian peafowl has been kept by humans in semidomestication for over 2,000 years and is better known as the peacock (male) and peahen (female).

Because the Indian peafowl, found naturally in India, Sri Lanka and parts of Pakistan, is so well known, it hardly needs description. The peacock's head, neck and breast are a glossy blue, relieved only by white patches, one above and one below the eye. The head also bears a crest made up of a bunch of divergent, brushlike feathers tipped with blue. The body is drab by comparison, being grayish barred with brown on the back and brown wings. However, the real beauty of the peacock is its spectacular green train, often called its tail, although it is made up of greatly elongated tail coverts, rather than of tail feathers. Including its train, a peacock measures up to 7½ feet (2.3 m) long, of which nearly two-thirds is the train. The peahen is less showy than the peacock. She has a whitish face and throat, brown crown and hind neck and a metallic green upper breast. She lacks the sweeping train of the male, but has the same fanlike crest.

Both the male and female of the green peafowl are green with bronze mottlings from the crest to the tail. The wings are turquoise and black, and the entire plumage has a metallic appearance. The crest is smaller than in the Indian peafowl. Albino, pied and black-shouldered mutations have appeared, but these are soon lost by crossing with pure stock. The Indian and green peafowls hybridize in captivity, the offspring looking like the Indian parents.

The Congo peafowl, not discovered until 1936, lives in deep forests. The male is blue and green with a short broad tail, a patch of naked red skin on the neck and a double crest of black and white. The female is chestnut and coppery green. This species has not been seen since 1982 and is likely to be endangered.

INDIAN PEAFOWL

CLASS **Aves**

ORDER **Galliformes**

FAMILY **Phasianidae**

GENUS AND SPECIES ***Pavo cristatus***

ALTERNATIVE NAMES
Peacock (male); peahen (female); blue peafowl, common peafowl (both sexes)

WEIGHT
Male: 8⅘–13¼ lb. (4–6 kg); female: 6–8⅘ lb. (2.75–4 kg)

LENGTH
Head to tail: male, 6–7½ ft. (1.8–2.3 m); female, 3–3¼ ft. (0.9–1 m)

DISTINCTIVE FEATURES
Very large size. Male: fanlike crest; glossy blue crown, neck and breast; metallic green back; spectacular green train. Female: lacks train; whitish face and throat; brown crown, hind neck and back; metallic green upper breast; white belly.

DIET
Mainly seeds, fallen fruits and insects; rarely small reptiles and rodents; also scraps

BREEDING
Age at first breeding: 2–3 years; breeding season: eggs laid January–September; number of eggs: 3 to 6; incubation period: 28–30 days; breeding interval: 1 year

LIFE SPAN
Up to 10 years or more

HABITAT
Open deciduous woodland, forest and orchards, often near streams

DISTRIBUTION
Pakistan east of Indus River; most of India and Sri Lanka

STATUS
Locally common

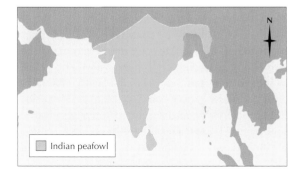

Screeching calls

Indian peafowl feed on the ground and roost in the trees, bugling loudly in the late afternoon as they go up branch by branch. Their loud screeching call can also be heard intermittently after dark. Where they are persecuted, they soon react by keeping out of sight in the forests, and when they do so, their plumage serves as effective camouflage. Where they are undisturbed, peafowl readily live around human settlements, attaching themselves to particular buildings. This is a marked feature of their behavior in semidomestication, and it is linked with an unusual regularity in their habits generally. Indian peafowl not only keep to the same roost and visit the same feeding grounds and sunbathing places day after day, but the male also has a chosen place for displaying.

Nothing edible refused

Peafowl eat grain and seeds, fruits, insects and, in gardens, bread and other scraps. They seem to be ready to try almost anything edible, plant or animal, and these birds have been seen to kill and eat small mammals such as mice, and even small snakes. They will, like the barnyard chicken, snap flies from the air or chase butterflies.

The fan dance

The peacock's train is described in the ornithological literature as made up of elongated tail coverts, partly decomposed and tipped with ocelli, or eyespots. The "decomposition" means the barbules of the feathers do not interlock, having lost their hooks, making them look ragged. Nonetheless, the train is still a spectacular sight when spread in a fan, showing off its

The preferred natural habitat of the Indian peafowl is deciduous woodland, near streams, although it can also be found in orchards and open forest. This photograph shows a pair of peacocks in Kanha National Park, India.

Indian peafowl, such as this peahen, have been semidomesticated for over 2,000 years. Often the birds are protected for religious, sentimental or aesthetic reasons, and in these cases they become quite tame.

greenish, shimmering colors and the beauty of the eyespots. The peacock displays his fan to the peahen or to any other passing bird, even to humans, sometimes wheeling around and displaying the back view of the fan as well. He often appears to posture and strut while displaying, and it is from this that the phrase "proud as a peacock" is derived. However, such posturing is often due to the peacock trying to keep his balance as the wind catches this enormous fan. From time to time he quivers the fan or rattles its quills. The male's train grows in length as the bird ages, at least until he is in his fifth or sixth year.

The whole displaying effect seems to be largely lost on the four or five hens in the peacock's harem, but eventually mating takes place. Nor is the display the prerogative of the peacock alone: the peahen, even the peachicks, may be seen to posture with their undersized tail coverts spread.

Foster parents

Where there are many ground predators, the Indian peafowl's nest is sometimes built in a tree. More usually, however, it is a shallow scrape in the ground in which the three to six eggs are laid. Egg laying takes place from January to March across most of southern India, but may be as late as September elsewhere. The eggs hatch after

28–30 days and the hen feeds the chicks from her bill, at which the chicks instinctively peck. In domestication peafowl eggs are sometimes put under barnyard hens. The chicks from these have to learn to pick up food from the ground. Orphaned chicks placed in the care of an immature peahen will peck at her beak. After a while this stimulates her to feed them, and she will later take them under her wings.

Changing fortunes

The peafowl is venerated in parts of its native range. Elsewhere it is killed for food. Alexander the Great introduced it to the ancient Greeks, who made it sacred to Hera, queen of the heavens. It later became associated with Juno, Hera's Roman counterpart, as a regal bird, but by the first century B.C.E. roast peacock had become fashionable at banquets and was a symbol of extravagant feasting. The fashion spread north and west into medieval Europe.

Peacock feathers, by contrast, have at all times been prized, and in 17th-century Scandinavia the birds were reared solely for their aesthetic value, as indeed they are today, being ornamental additions to many private and public gardens. The Indian peafowl is also protected for religious and sentimental reasons in parts of India, and in these regions it has become tame. Where it is still hunted, it remains shy and secretive.

PEARLFISH

Pearlfish are so named because they are often found embalmed in pearls in oyster shells and because their bodies have a pearly luster. Species may be free living (genus *Echiodon*), commensal (*Carapus* and *Onuxodon*) or parasitic (*Encheliophis* and *Jordanicus*). Commensalism is a relationship between two kinds of organisms in which one obtains food or other benefits from the other without damaging or benefiting it. The pearlfish coexist with shallow-water invertebrates such as holothurians (sea cucumbers), bivalve mollusks and starfish. Many pearlfish live in coral-dominated communities.

Transparent and eel-like

The 24 species of pearlfish are thin and eel-like, usually only a few inches long. The longest, *Echiodon drummondi* and *Carapus bermudensis*, can reach 12 inches (30 cm). Pearlfish have scaleless, rounded or deep bodies. The tail tapers almost to a point. The dorsal fin starts just behind the head and continues around the tail to join the long anal fin. The pectoral fins are small and there are no pelvic fins. The body is transparent, so the internal organs and the backbone are visible.

Most species of pearlfish live in the world's tropical and subtropical seas, with a few in warm temperate seas. They are found in waters over the continental shelf and the adjacent slope, from coastal shallows out to depths of approximately 6,560 feet (2,000 m).

Remarkable partnerships

Adult pearlfish shelter in cracks and crevices in rocks, inside the shells of bivalve mollusks or inside the bodies of sea squirts, sea urchins, starfish and sea cucumbers. Most species of pearlfish come out to feed, but a few live permanently and parasitically inside the body of another animal. Pearlfish are occasionally caught in nets, which suggests that some of them may spend a larger proportion of their time swimming freely than hiding in crevices.

The hosts used by pearlfish are varied but are often specific to a particular species. For example, *Encheliophis homei*, an uncommon pearlfish of the Malaysian Archipelago, readily inhabits the body cavity and respiratory trees of some holothurians, especially *Stichopus chloronotus* and *Holothuria aspus*. It is apparently in competition with another pearlfish, *Jordanicus gracilis*, for its preferred host, *H. aspus*. By contrast, the pearlfish *Carapus bermudensis* of the Caribbean uses only three species of sea cucumbers. It spends only part of its time within the host. Other pearlfish species, such as *C. acus* of

Most pearlfish have a close relationship with invertebrates, particularly sea cucumbers. The fish use these hosts as daytime shelters, entering and exiting through their vents (below). A few species of pearlfish live permanently inside.

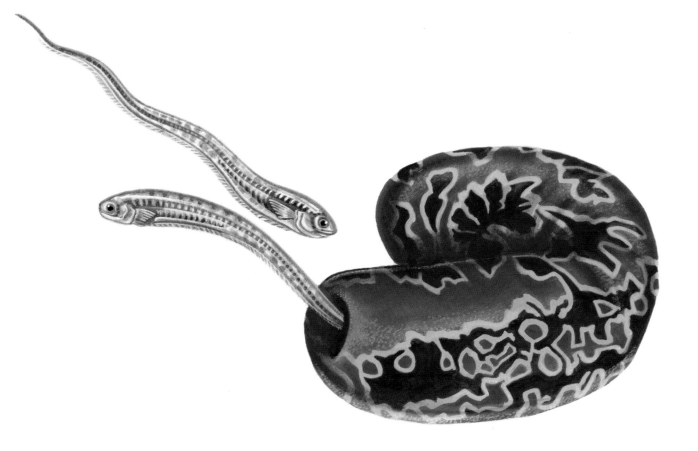

the Mediterranean, live permanently inside such species as *H. tubulosa* and *S. regalis*. These pearlfish are parasites.

Studies reveal that a pearlfish that normally uses a sea cucumber for a host will enter an artificial sea cucumber so long as water is flowing out of it and some mucus from a live sea cucumber has been added to the aquarium water. A sea cucumber breathes by drawing water in through its vent, passing it through its gills and then driving it out again. Young pearlfish swim in headfirst, using the head to keep the sea cucumber's vent open while the tail is being inserted. Should the sea cucumber try to close its vent before the fish is safely inside, a pearlfish may twist its long tail like a corkscrew to insinuate itself inside. More than one pearlfish may live in the same sea cucumber.

Dining out and in

For the most part pearlfish shelter by day and come out at night to feed on copepods (a large group of tiny crustaceans), small shrimps, small crabs and probably other small invertebrates. The parasitic species of pearlfish feed on the reproductive organs of their hosts and possibly other internal organs. This may not be as bad as it appears for the sea cucumbers, which have the habit of casting out their internal organs in moments of crisis and growing a new set, so presumably they can regrow any parts that are nibbled away by the pearlfish.

Packets of eggs

At breeding time the female pearlfish lay eggs in sticky, roughly cylindrical masses 2–3 inches (5–7.5 cm) long. After hatching, each baby fish passes through four stages, with changes in the body shape, coloration and behavior occurring at each stage. In *C. acus* the first or vexillifer larva is very slender and has a long lobe on its back. This is not a fin but a flaglike outgrowth of the body (*vexillifer* means "standard-bearer"). As the larva grows in size, the flag becomes small, disappearing when the young fish is 2½ inches (6 cm) long. All this time it has lived among the plankton. The next stage, the tenuis or slender larva, has a larger head relative to the rest of the body, and it begins to grow long teeth. It must now enter a sea cucumber. It cannot transfer from one sea cucumber to another and it cannot survive outside. When it has reached a length of 8 inches (20 cm), the tenuis larva changes to a juvenile. This is much shorter, and can leave its host for a while each day, but it feeds on the gills and the reproductive organs of the sea cucumber. The fourth and final stage is the adult. As the adult pearlfish gets older, the sea cucumber is used only as a temporary day refuge.

PEARLFISH

CLASS	**Osteichthyes**
ORDER	**Ophidiiformes**
FAMILY	**Carapidae**

GENUS AND SPECIES **24 species, including** *Carapus acus; C. homei; C. bermudensis; Jordanicus gracilis;* **and** *Echiodon drummondi*

LENGTH
Up to 8 in. (20 cm)

DISTINCTIVE FEATURES
Eel-like body tapering from head to pointed tail; large jaws extending back past eye; dorsal and anal fins run along body and unite at tail; no pelvic or tail fins; scaleless and transparent with reddish spots

DIET
Free-swimming species: small invertebrates such as shrimps and crabs; parasitic species: mainly reproductive organs of host itself

BREEDING
Varies according to species

LIFE SPAN
Not known

HABITAT
Coastal waters, in close association with hosts such as sea cucumbers, sea squirts, sea urchins, starfish and bivalve mollusks

DISTRIBUTION
Most species: tropical and subtropical seas

STATUS
Not known

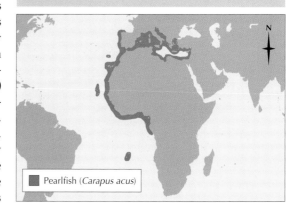
Pearlfish (*Carapus acus*)

Nothing definite is known about the predators of pearlfish, but when the fish have started to shelter, whether in rock crevices or in living animals, their daily risks must be much reduced. No doubt some fall prey to larger predatory fish when they are out feeding.

PECCARY

PECCARIES ARE THE American equivalents of the wild swine or pigs (family Suidae) of Asia, Africa and Europe, which they superficially resemble. Peccaries are smaller than the true swine, however, and differ in other important details, so they are placed in a separate family, the Tayassuidae.

Peccaries have long, slim legs and small hooves. They have only three toes on the hind feet, whereas the true swine have four. The tail is vestigial (imperfectly developed), and the body is covered with thick bristly hairs that form a slight mane on the neck. When a peccary becomes agitated, the hairs along the spine are raised, exposing a scent gland on the lower back. This gland exudes an unpleasant musky odor that is detectable at some distance. Peccaries have more complex stomachs and fewer teeth than do true swine, and their short, sharp upper tusks grow downward instead of upward.

There are three species of peccaries. The collared peccary, *Pecari tajacu*, is the most widespread and is found in deserts, woodland and rain forests from the southern borders of the United States to Argentina. The white-lipped peccary, *Tayassu pecari*, ranges from Paraguay to Mexico. The Chacoan peccary, *Catagonus wagneri*, lives in the Gran Chaco, a huge expanse of thorn bush and thorn forest in southeast Bolivia, Paraguay and northern Argentina.

Of the three species, the collared peccary is the smallest, standing 20 inches (51 cm) at the shoulder, with a maximum length of about 37 inches (94 cm). It weighs up to 65 pounds (30 kg). The coarse hair is black mixed with white, so the effect is grayish. The animal owes its name to the narrow, semicircular collar of lighter hair on the shoulders. Alternative names for the collared peccary are musk hog or javelina, the latter a reference to its spearlike tusks.

A young collared peccary keeps close to its mother in the Tambopata Reserve, Peru. Usually a female breeds only once a year, but she may mate a second time if she loses her first litter of the season.

A collared peccary forages for food. Although the adults are unpredictable, peccaries have been tamed and kept as pets.

PECCARIES

CLASS	**Mammalia**
ORDER	**Artiodactyla**
FAMILY	**Tayassuidae**

GENUS AND SPECIES **Chacoan peccary, *Catagonus wagneri*; collared peccary, *Pecari tajacu*; white-lipped peccary, *Tayassu pecari***

ALTERNATIVE NAMES
Collared peccary: javelina; musk hog; Chacoan peccary: tagua

WEIGHT
31–110 lb. (14–50 kg)

LENGTH
Head and body: 29½–44 in. (75–112 cm); shoulder height: 17⅓–27 in. (44–69 cm)

DISTINCTIVE FEATURES
Piglike but with slimmer legs and small hooves; bristly coat with mane of longer hairs along back; small tusks (upper canines)

DIET
Mainly roots, bulbs, seeds and fruits; also grubs, worms, carrion, snakes and small mammals; cacti (Chacoan peccary only)

BREEDING
Age at first breeding: 1–2 years; breeding season: February–March; number of young: 1 to 4; gestation period: 142–148 days; breeding interval: 1 year

LIFE SPAN
Up to 24 years in captivity

HABITAT
Woodland, rain forest, thorn forest, bush, steppe and desert scrub

DISTRIBUTION
Collared peccary: southern U.S. south to northern Argentina

STATUS
Collared and white-lipped peccaries: common; Chacoan peccary: endangered

Collared peccary

Back from the dead

The white-lipped peccary is dark reddish brown to black in color with, as the name suggests, an area of white around the mouth. It can measure as much as 41 inches (104 cm) long and stands 24 inches (61 cm) at the shoulder. The Chacoan peccary, also known as the tagua, is easily the largest of the peccaries, measuring 35½–44 inches (90–112 cm) in length. In color it is a grizzled mixture of gray brown, black and white, with a black stripe along its back and a white collar. It has blackish legs. Zoologists believed the Chacoan peccary had become extinct, but local people knew it still existed, and in 1972 it was finally rediscovered by science, being documented three years later.

Safety in numbers

The speed and agility of peccaries, which move with a fast, running gait when chased, combined with a group defense system, protect them from dogs, coyotes (*Canis latrans*) and even bobcats (*Felis rufa*). Although they are not normally aggressive to humans, an entire herd may counterattack if a member is wounded or chased. Peccaries are hunted for their hides and meat, and when at bay, the herds are described as standing in close formation, champing their tusks and making charges at their hunters.

All three peccary species are gregarious, but whereas the collared peccary associates in groups of 5 to 15 and the Chacoan peccary in bands of 2 to 10, herds of white-lipped peccaries are much larger, containing up to 100 or more individuals. Peccary groups contain males and females of all ages. Herds do not appear to move over large areas. The territorial range of the collared peccary, for example, is usually only about 3 square miles (7.8 sq km).

Peccary herds have no apparent leader, and it has been suggested that the musk gland plays an important part in coordinating group movements. Peccaries of the same herd practice mutual grooming, in the course of which they rub their throats and shoulders on each other's musk glands, a habit known as smell-sharing behavior. In the normal course of events, secretions from these glands are also transferred to low branches or bushes along paths frequented by the herds. In this way, the home territory, as well as each fellow member of a herd, is made instantly recognizable by smell. Peccaries also have the habit of pawing sand against themselves, which is probably a cleaning action.

Rooting for food

Peccaries are most active in the cooler hours of the day and at night. Their usual resting place is in a thicket or under a large boulder. They also readily take to the abandoned burrows of other animals. Peccaries' eyesight is not good and their hearing only fair, but they have a keen sense of smell, being able, for instance, to locate an edible bulb 4 inches (10 cm) or so underground. The collared peccary is mainly vegetarian and uses its piglike snout to root for fruits, berries and bulbs. However, it also eats grubs, occasionally small vertebrates and even snakes. It appears to be immune to rattlesnake venom. The white-lipped peccary is more omnivorous, living on carrion, worms and insects, as well as on a variety of fruits and roots, and is reputed to hunt larger prey. It is never far from running water. The Chacoan peccary's preferred food is cacti, but it too feeds on fruits and roots.

Short infancy

Very little is known of the breeding habits of the white-lipped peccary. A collared peccary litter, however, consists of 1 to 4 young, as does a Chacoan peccary litter. Female collared peccaries may breed twice in the same year. Gestation is 142–148 days, and the young are born in a burrow, hollow log or cave. Their color is quite different from an adult's, being reddish with a dark stripe down the back. They are able to run when a few hours old and are weaned at 6–8 weeks. Life expectancy for a collared peccary is up to 24 years in captivity.

Collared peccaries form herds of 5 to 15 individuals. This species ranges north into the United States, where it can be found in the Chihuahuan and Sonoran deserts of southwestern Texas, in southern New Mexico and in southwestern Arizona.

PEDIPALPI

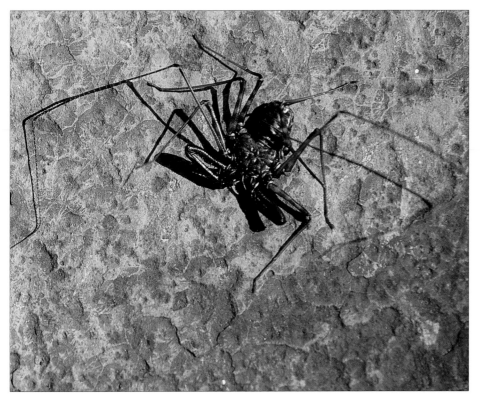

Whip scorpions have neither stings nor venomous bites, and members of the order Amblypygi, such as Damon variegatus (above), are harmless to humans.

THE PEDIPALPI ARE AN assemblage of animals now regarded as belonging to three separate orders in the class Arachnida. The members of two of these groups are known as whip scorpions, but for different reasons. The whip of one group, the Uropygi, is a long, thin, tactile tail at the end of the abdomen, while the other whip scorpions, the Amblypygi, have two whips in the form of a pair of very long, flexible legs. The members of the third group, the Schizomida, are small and retiring. All three groups use the first pair of legs as feelers rather than for walking, but in the Amblypygi the length is particularly exaggerated. In all three orders, the body is in two parts, a cephalothorax (combined head and thorax) and an abdomen of 12 segments. These vary from less than ½ inch to about 2¾ inches (1–7 cm) long, excluding the tail of the Uropygi. The cephalothorax bears the legs and, at the front, a short pair of anterior appendages called chelicerae followed by the well-developed pair of sensory appendages, known as pedipalps, after which these animals are named.

Powerful pedipalps

The pedipalps of the Uropygi (literally, "tail-rumps") are armed with claws for the capture of cockroaches, grasshoppers and other insects, as well as slugs and worms, or even small frogs and toads in the case of the largest species, *Mastigoproctus giganteus*, of Mexico and the southern United States. At the base of each pedipalp is a large semicircular toothed structure, used for crushing the prey. There are 8 to 12 weak eyes arranged in three groups on the cephalothorax, but the first pair of legs are more important as sense organs. The Uropygi are nocturnal hunters, hiding during the day under logs and stones or in burrows. *M. giganteus* spends several days digging its burrow with its pedipalps. When it is finished, the tunnel may be 4 inches (10 cm) long and it is here that the prey is usually eaten. There are between 100 and 130 species of Uropygi living in southern North America and northeastern South America, India, Peninsular Malaysia and eastern Asia.

The members of the small group Schizomida (meaning "split middles"), are ¼ inch (0.6 cm) or less long, with at most only a knob for a tail. The carapace of the cephalothorax, divided in this order alone into three parts, carries only one pair of eyes. The pedipalps end in spines instead of claws and they move up and down instead of sideways, as in the Uropygi. There are three genera and the group occurs sporadically in tropical regions. The Schizomida feed at night, probably on insects, and hide by day, though not in any fixed home. When disturbed, they escape by a quick backward leap and can run quickly.

The tailless Amblypygi, (literally, "blunt-rumps") have very flattened bodies that are suitable for getting through narrow cracks, and the two halves are joined by a slender stalklike abdominal segment. There is one pair of median eyes and three pairs of lateral eyes, as in some of the Uropygi. The pedipalps are spined, powerful and sometimes very long, and each ends in a movable hook. There are about 130 species of Amblypygi, ranging in length from less than ⅓ inch to nearly 2½ inches (0.5–6 cm). They inhabit humid tropical and subtropical regions: the southern half of Africa, America, India, Peninsular Malaysia, Borneo and New Guinea.

Nocturnal like the other pedipalpi groups, the Amblypygi cling by day to the undersurfaces of rock crevices, logs and stones. Some species, with less need for dampness, have become commensal with humans, and in some parts of the world few houses are without them. When they are exposed to light, their first reaction is to freeze, but they will run fast if touched. They usually walk sideways and, as when at rest, continually search around with the tips of their long legs. One of the three Ampblypygi families consists mostly of small cave-dwellers and these, unlike the others, can run around under the ceil-ings of caves or even up polished glass. The prey, mostly insects of various kinds, are held in the spiny pedipalps while the chelicerae remove pieces for chewing.

Courtship dances

During courtship among the Uropygi the courting male holds the long front legs of the female in his pedipalps and chelicerae and walks backward with his mate following. The male deposits a spermatophore (sperm parcel) on the ground and, having grasped the female with his pedipalps, pulls her over the spermatophore, which is then coaxed into her genital opening. The pregnant female digs a burrow where she stays several weeks and lays, in one species at least, 20 to 35 yellowish eggs that are retained in a transparent membrane under her abdomen. The young cut their way out of this by means of special spines on their legs and cling to the upper side of the mother's abdomen or to the bases of her back legs until they reach the adult form, at the first molt. They then leave the mother, the strength of which is nearly spent, and become adult after three more yearly molts.

In the single member of the Schizomida so far studied by scientists, the mating pair promenade, with the female holding the end of the male's abdomen with her chelicerae. He then deposits a spermatophore and cements it to the ground. From the top of this, the female gathers the sperm. Later she builds a little nest with cemented walls under the soil.

The Amblypygi court at night, with much tapping of the front legs and threatening with the pedipalps, but no grasping. The male deposits a slender, transparent spermatophore on the ground while facing away from the female, then turns toward her and loads it with sperm. As he steps back and quivers, she collects the sperm, leaving the spermatophore for him to eat. She also carries her eggs in a sac under her abdomen.

PEDIPALPI

PHYLUM **Arthropoda**

CLASS **Arachnida**

ORDER **Uropygi; Amblypygi; Schizomida**

FAMILY **Uropygi: 2 families; Amblypygi: 3 families; Schizomida: 1 family**

GENUS AND SPECIES **130 species in both Uropygi and Amblypygi; few in Schizomida**

ALTERNATIVE NAMES
Whip scorpion, vinegaroon (Uropygi); tailless whip scorpion, whip spider (Amblypygi); short-tailed whip scorpion (Schizomida)

LENGTH
Up to ⅓–3¼ in. (0.5–8 cm), depending on species and sex

DISTINCTIVE FEATURES
Body divided into cephalothorax (combined head and thorax) and abdomen; 4 pairs of legs. Whip scorpions: 1 pair of large pedipalps (sensory appendages) at front of cephalothorax; long, whiplike tail at rear of abdomen. Whip spiders: elongated, whiplike front legs used for sensory perception; clawlike, spiny pedipalps; no tail. Tailless whip scorpions: generally smaller than other groups; no abdominal tail.

DIET
Most species: other invertebrates; largest Uropygi species: also small frogs and toads

BREEDING
Number of eggs: 20 to 35 (Uropygi species)

LIFE SPAN
Some species may live for many years

HABITAT
Humid crevices in caves or under rocks, leaf litter or bark; many species dig burrows

DISTRIBUTION
Most species in Tropics and subtropics

STATUS
Generally common

Uropygi, such as M. giganteus (below), can discharge an acidic secretion from glands near the anus. The secretion may be up to 84 percent acetic acid, or vinegar, hence the animals' common name "vinegaroon."

PELICAN

Brown pelicans nest in colonies in trees, bushes or on the ground. Tree nests are made of reeds, grasses, straw and sticks.

THERE ARE SEVEN SPECIES of pelicans, all of which belong to the same genus. Two of these species occur in the New World and five in the Old World, distributed over the tropical and warm temperate parts of the globe. Both sexes are alike and all have massive bodies, supported on short legs with strong webbed feet. They have long necks, large heads and a thick, tough plumage. Pelicans are among the largest living birds. The largest of all pelican species is the silvery white Dalmatian pelican, *Pelecanus crispus*, which reaches 5¼–6 feet (1.6–1.8 m) in length, with a wingspan of 10¼–11½ feet (3.1–3.5 m). It weighs 22–26½ pounds (10–12 kg). The most conspicuous feature is the enormous bill: the upper part is flattened, and the lower part carries a pouch, known as a gular pouch, that can be distended considerably. The pouch of the brown pelican, *P. occidentalis*, can hold about 3 gallons (13.5 l) of water and is used as a dip net for catching fish rather than to store food.

Apart from the brown pelican, in the majority of pelican species the adult plumage is mainly white, tinged with pink in the breeding season in some species such as the pink-backed pelican, *P. rufescens*, of Africa. The primary feathers are black or dark. Some species have crests and in some there is yellow, orange or red on the bill, pouch and bare parts of the face.

Only one marine species

The brown pelican is the smallest member of the pelican family, with a wingspan of up to 9¼ feet (2.8 m) and weighing 4½–8¾ pounds (2–4 kg). It has a white head with a yellow tinge to it. In the breeding season the neck turns a rich brown with a white stripe running down each side. The wings and underparts are dark brown. The brown pelican is a seabird, but does not venture far from the shore. It is found along the southern Atlantic and Gulf coasts of North America through the Caribbean to the Guianas. Along the Pacific it ranges from central California to Chile, with one population on the Galapagos Islands. The brown pelican is the only truly marine pelican. The remaining pelican species are found on lakes, estuaries, river deltas and lagoons. The other New World species is the American white pelican, *P. erythrorhynchos*, which breeds on inland lakes from western Canada to southern Texas.

In the Old World there are pelicans in Africa, southern Asia, the Philippines and Australia, and in southeastern Europe there are isolated colonies of the Dalmatian pelican, which ranges eastward into Central Asia, visiting Egypt and northern India in winter. The breeding range of this species has contracted and today only about 100 pairs nest in Europe, with about 1,400 to 2,000 pairs in Turkey, Central Asia and Mongolia. It winters from the Balkans east through Iran and the Persian Gulf to Pakistan and India. Classed as vulnerable by the World Conservation Union (I.U.C.N.), the Dalmatian pelican has become extremely localized.

BROWN PELICAN

CLASS **Aves**

ORDER **Pelecaniformes**

FAMILY **Pelecanidae**

GENUS AND SPECIES *Pelecanus occidentalis*

ALTERNATIVE NAMES
Chilean pelican; Peruvian pelican

LENGTH
Head to tail: 4–4½ ft. (1.2–1.4 m); wingspan: up to 9¼ ft. (2.8 m)

DISTINCTIVE FEATURES
Huge, long bill with extendible pouch suspended below; large head; broad, heavy body; long, broad wings; webbed feet. Adult: grayish brown to brown-black overall; blackish throat patch; golden yellow head and neck sides (whiter when not breeding).

DIET
Mainly fish; some crustaceans

BREEDING
Age at first breeding: 3–5 years; breeding season: mainly March–April (U.S. Pacific coast), December–June (U.S. Gulf coast), all year (Chile and Peru); number of eggs: 2 or 3; incubation period: 30–32 days; fledging period: 75–84 days; breeding interval: 1 year

LIFE SPAN
Probably up to 40 years

HABITAT
Coastal waters; also on inland fresh waters. Nests on slopes, islands and sandy beaches.

DISTRIBUTION
Coasts from California south to Chile, and Maryland south to northern South America

STATUS
Locally common (Florida, California, Peru); uncommon (rest of U.S.); varies elsewhere

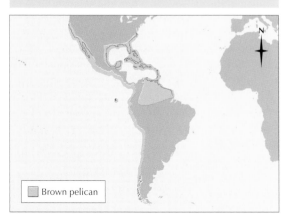

Brown pelican

A pair of Australian pelicans, Pelecanus conspicillatus. *The huge pouch beneath the pelican's bill has a capacity two or three times greater than that of the bird's stomach.*

Pelicans feed almost entirely on fish but a few crustaceans are also taken. White pelicans fish while floating on the surface or wading about in the shallows. They thrust their heads under the water, using their pouches as dip nets to catch the fish.

Community breeding

Pelicans are very sociable and all the species nest in large colonies sometimes consisting of tens of thousands of birds. Most of the white species breed on isolated islands in large inland lakes, usually making their nests on the ground but occasionally they nest in low trees. On the ground the nest is sometimes just a depression scooped out of the earth. The brown pelican, which breeds on small islands on the coast, makes a loose nest of sticks in mangrove trees and low shrubs or sometimes on the ground.

In all species the breeding season varies from place to place and from year to year. In some tropical areas, such as parts of the Caribbean and Chile and Peru in the case of the brown pelican, the birds may even breed throughout the year. The eggs are chalky white in appearance, and the number laid varies according to species. Both parents help to incubate the eggs for 29–35 days, again according to species. Such a period is fairly

Great white pelicans fish cooperatively. They form a line across the water and beat the surface, driving small fish into shallow water to scoop them up.

short for such a large bird. The chicks are born naked and blind but quickly grow a soft white down. Both parents feed the young, at first dribbling regurgitated food out of the ends of their bills into the chicks' open mouths, but after a few days the chicks are strong enough to stick their heads into their parents' pouches to get the food. Before the chicks are 2 weeks old they leave the nest and form noisy juvenile groups, but the parents continue to feed them for some time. The young mature slowly, only acquiring adult plumage after several years. They seldom breed until they are 4 years old. Pelicans are long-lived birds and have a low adult mortality rate: the accepted record for pelican longevity is 40 years.

Many hazards for the young

Mature pelicans have few natural enemies. Sometimes they may be killed by sea lions in the Pacific or occasionally eaten by sharks, but among the young mortality is very high. About 70 percent of young pelicans die within their first year, a figure that goes some way toward counterbalancing the pelican's long reproductive lifespan. When the young birds congregate after leaving the nest, many fall from trees or get caught in the branches or even trampled by clumsy adults. The adult birds do little to protect their young, and sometimes entire nesting colonies are wiped out by predatory animals. Fishers have been known to destroy colonies of pelicans to prevent them from taking too

many fish. Today pelican colonies are commonly endangered by marsh drainage, water pollution, habitat loss or from lakes drying up as the result of large water schemes. Variable fish stocks also contribute toward a high chick mortality.

Strong in flight

When a pelican has managed, after much effort and flapping to become airborne, it is a strong and graceful flier, and it is no less graceful in the water. With legs up, head well back on the shoulders and its large bill resting on the front of the neck, it can sail through the air with little effort.

Many pelican species are part-migratory, or nomadic, and regularly move feeding grounds and breeding grounds. The ability to fly steadily over some distance is essential to them. Pelicans fly at about 26 miles per hour (42 km/h) and there is an authentic record of their having maintained this speed for 8 miles (13 km), so it seems they also have the quality of endurance in flight. There is one record of the great white pelican, *P. onocrotalus*, having achieved 51 miles per hour (82 km/h).

Pelicans regularly fly in formation, either in single file or in V-formation, all members of the flight beating their wings in perfect unison. They also use thermal currents, in the manner of vultures, soaring in spirals to a great height, perhaps as high as 8,000 feet (2,400 m), where by alternately flapping and gliding they may circle for hours.

PENGUINS

Among the toughest and most specialized of all birds, penguins survive in a greater range of temperatures than any other bird family. On its Antarctic breeding grounds the emperor penguin, *Aptenodytes forsteri*, endures fierce blizzards during which the air temperature may plummet to -70° F (-55° C), and thousands of miles to the north the Galapagos penguin, *Spheniscus mendiculus*, inhabits warm equatorial waters off the remote Galapagos Islands. In the scorching heat of the South African summer, small offshore islands are home to the jackass penguin, *S. demesus*, while the yellow-eyed penguin, *Megadyptes antipodes*, nests among vegetation and tree roots on the floor of temperate rain forests on South Island, New Zealand.

Penguins are flightless seabirds with modified wings, no flight feathers and short, strong legs placed far back on the body. They sacrificed the ability to fly in order to become supremely efficient underwater hunters, capable of swimming much faster and diving far deeper than other waterbirds. Penguins range in size from about 14 inches (35 cm) tall and approximately 2 pounds (1 kg) in weight, in the little blue or fairy penguin, *Eudyptula minor*, to 45 inches (1.15 m) and 42–100 pounds (19–45 kg) in the emperor penguin.

Classification

Penguins belong to the order Sphenisciformes, which contains a single family (Spheniscidae) found only in the Southern Hemisphere. There are six genera of penguins, and most authorities currently recognize 17 species. However, some ornithologists consider a New Zealand subspecies of the little blue penguin (which is mainly found around the southern coasts of Australia) to be a full species in its own right, usually known as the white-flippered penguin, *E. albosignata*.

The fossil record of penguins begins in the late Eocene, about 45 million years ago, with most fossil specimens coming from New Zealand, Patagonia, the Antarctic and southern Australia. At least two of these ancient penguins were giants measuring 5¼ feet (1.6 m) tall and weighing about 300 pounds (135 kg). It is not clear when penguins lost the ability to fly, although they presumably passed through an evolutionary stage in which they could both fly and swim underwater. Penguins do not have any close relatives, but they may be distantly related to the albatrosses and petrels of the order Procellariiformes.

CLASSIFICATION
CLASS Aves
ORDER Sphenisciformes
FAMILY Spheniscidae
GENUS Aptenodytes Pygoscelis Eudyptes Megadyptes Eudyptula Spheniscus
NUMBER OF SPECIES 17

A gentoo penguin, Pygoscelis papua, *feeding its chick by regurgitation. Chicks grow fast on their protein-rich diet.*

Origin of the name

The name penguin is derived from the Latin name of a now-extinct seabird, the great auk, *Pinguinus impennis*. The great auk was the only flightless bird found in the Northern Hemisphere during historic times. Once an abundant species, the last pair was shot in 1844. The surviving species of auks (family Alcidae), including the little auk or dovekie, puffins, murrelets, razorbill and guillemots or murres, all resemble penguins, although the two groups are entirely unrelated. The auks and penguins are an excellent example of convergent evolution, in which two types of animal have evolved similar adaptations and lifestyles in complete isolation from one another. The auks are discussed elsewhere, under "Seabirds."

Mainly black and white

Most species of penguins have blue-black or blue-gray upperparts and white underparts, often with lines of black across the upper breast or spots of white on the head. Color is

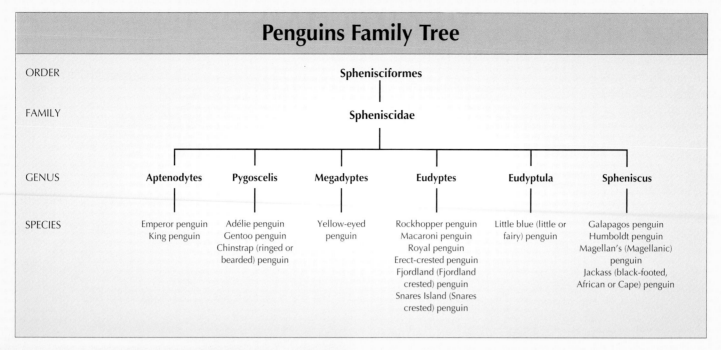

Penguins Family Tree

ORDER	Sphenisciformes					
FAMILY	Spheniscidae					
GENUS	Aptenodytes	Pygoscelis	Megadyptes	Eudyptes	Eudyptula	Spheniscus
SPECIES	Emperor penguin King penguin	Adélie penguin Gentoo penguin Chinstrap (ringed or bearded) penguin	Yellow-eyed penguin	Rockhopper penguin Macaroni penguin Royal penguin Erect-crested penguin Fjordland (Fjordland crested) penguin Snares Island (Snares crested) penguin	Little blue (little or fairy) penguin	Galapagos penguin Humboldt penguin Magellan's (Magellanic) penguin Jackass (black-footed, African or Cape) penguin

A rockhopper penguin, Falkland Islands. All penguins in the genus Eudyptes have bright yellow or orange brow tufts.

limited to red, pink or orange bills or feet; red or yellow eyes in some species, such as the chinstrap penguin, *Pygoscelis antarctica*; yellow or orange brow tufts in the six species of the genus *Eudyptes*; and patches of orange-yellow on the head, neck and breast in the very similar emperor penguin and king penguin, *Aptenodytes patagonicus*. The main chick plumages are gray or brown, sometimes with whitish underparts.

Physical adaptations

The squat appearance of penguins, together with their upright stance on land and comical, waddling way of walking, has endeared them to humans. Penguins' webbed, rear-set legs are ideal for steering underwater but make movement on land difficult. Emperor penguins and Adélie penguins, *Pygoscelis adeliae*, often toboggan down icy slopes on their bellies because sliding along is quicker and uses less energy than walking; some species toboggan on sandy beaches. In addition to the feet, the short, stiff tail acts as an extra rudder when in the water. Propulsion underwater is provided by the powerful, narrow wings, which act rather like the flippers of dolphins, porpoises or seals. In effect, the wings of penguins have become paddles.

Penguins are superbly adapted for life in the water. Their bodies are sleek and streamlined, and a layer of fatty blubber under the skin insulates them from the cold water. Instead of having a plumage made up of various types of feathers, like other birds, the feathers of penguins are all alike. A penguin feather is short and stiff, and it lies flat against the body. Penguins also have comparatively solid bones, which means that they generally weigh only a little less than the water they displace, greatly reducing the energy required to dive. Like most birds, penguins have binocular vision (they look forward with both eyes), which enables them to judge distances accurately. However, light is not bent as much when it passes from water into the eye as when it passes from air. To compensate, the lens in a penguin eye is very strong. In consequence, penguins are relatively shortsighted on land.

Flying under the waves

The duration and depth of dives varies greatly between penguin species, but the emperor penguin can submerge for up to 18 minutes and reach at least 870 feet (265 m) below the surface, although 165 feet (50 m) is more usual. Even the small species are able to stay down for about 6 minutes and dive to almost 330 feet (100 m), which is much deeper than other diving birds such as auks, loons, cormorants and ducks.

The emperor penguin is one of the fastest swimmers, reaching speeds of up to 20 knots (36 km/h) in short bursts. However, swimming speeds average 3½–4½ knots (6–8 km/h). Swimming often involves porpoising, whereby the penguin breaks the surface with its momentum carrying it through the air for 3 feet (0.9 m) or so before re-entering the water. This may facilitate breathing while traveling fast near the water's surface, and it may also confuse potential predators.

Graceful hunters

Most penguins feed on crustaceans, squid and fish, which they catch in their long, razor-edged bills. The inside of the mouth and tongue are covered in backward-facing, fleshy knobs that help prevent the prey from slipping out of the mouth while water is being forced out from around it. Penguins swallow a considerable amount of seawater, and the excess salt has to be eliminated through specially enlarged tear glands.

Penguins, such as this Galapagos penguin, can power through the water at high speed, using their wings as paddles and their legs and tail as rudders. Their torpedo-shaped bodies and sleek feathers minimize drag.

Prey varies according to species and location. Penguins that feed close to the shore hunt mainly fish, whereas squid is a particularly important food for king, emperor and rockhopper (*Eudyptes chrysocome*) penguins. Krill (a small, shrimplike crustacean) forms the main diet of species in the genera *Pygoscelis* and *Eudyptes*, and is eaten by all penguins found in Antarctic waters.

Penguins hunt during the day, although the little blue penguin finds most of its food around sunrise and sunset. It is possible that penguins use echolocation to help detect prey, using a primitive sonar based on clicks produced by their swimming movement.

Breeding behavior

In general penguins are highly sociable, both at sea and on land. Many features of the life cycle vary with the body size of the species and its geographical distribution. Certain species, such as the jackass and little blue penguins, breed twice a year, but the majority of species breed only once each year and the king penguin twice in three years.

Most species begin breeding in the southern spring or summer, although some populations of the gentoo penguin also breed in winter. The breeding season of the emperor penguin starts in autumn, in March and April, timed so that the long developmental period will produce the young in midsummer, when their chances of survival are greatest.

Many types of vocal and visual displays are employed between the arrival of the penguins at the breeding colony and their departure. Various species have been described as trumpeting, croaking, cackling and cooing. The jackass penguin is so named because it sounds like a donkey braying. Ritualized postures include bowing and head and flipper waving, and penguins often present gifts of nesting material to their partner. Adélie penguins have the most elaborate displays of any penguin species. A penguin usually pairs up with the same mate year after year, unless the death of the latter forces it to find another partner. Penguins take a long time to reach sexual maturity: five or more years in most species.

One egg is laid by emperor and king penguins, and two (occasionally three) by all the others. The sexes share incubation duties except in the emperor penguin, in which they are undertaken by the male alone. Mortality of eggs and chicks is high, varying from 40 to 80 percent of all the eggs laid. Predators include skuas (genus *Catharacta*), sheathbills (*Chionis*) and gulls. Some penguins reduce the toll by nesting in burrows, visiting only at night. Burrow nesters include the yellow-eyed penguin and all *Spheniscus* species.

The colonies of emperor penguins are huge, containing 500 to 25,000 adults. Emperor penguins lay their eggs in the depths of the Antarctic winter to ensure that the young hatch in summer.

The adults take turns foraging at sea and brooding the chicks. In surface-nesting species, in which the chicks are more vulnerable to predators, the chicks eventually join a crèche (nursery group) containing up to 100 young. Crèches provide safety in numbers and sometimes are guarded by a few adults, while the other parents are at sea. Upon returning with food, each parent calls its chick from the crèche and is able to distinguish it from other chicks by voice and appearance. The period from hatching to independence varies from 2 months in the smallest species to 10–13 months in the king penguin.

Conservation

The populations of some species, such as the Adélie penguin, number several million, and the largest breeding colonies, on the subantarctic islands between 50°S and Antarctica, may contain hundreds of thousands of nesting adults. However, the disturbance of breeding colonies, oil spills and global warming are all reducing penguin populations. For example, 23,000 jackass penguins were coated with oil off the coast of South Africa when the oil tanker *Treasure* spilled its cargo in 2000. Climatic changes are influencing ocean currents, changing the movements of the prey on which penguins depend; this may be having an adverse affect on the ability of penguins to breed successfully. The World Conservation Union (I.U.C.N.) currently lists five species as vulnerable: the yellow-eyed, Galapagos, erect-crested (*Eudyptes sclateri*), Fjordland (*E. pachyrhynchus*) and Snares Island (*E. robustus*) penguins.

For particular species see:
- ADÉLIE PENGUIN • EMPEROR PENGUIN
- KING PENGUIN • YELLOW-EYED PENGUIN

PEPPERED MOTH

T HE PEPPERED MOTH BELONGS to the family Geometridae, or geo-meters as they are popularly known. These are night-flying moths that spend the day resting inconspicuously on tree trunks with the wings flat against the bark. The peppered moth has a wingspan of 1¾–2½ inches (4.5–6 cm) and flies in May and June in woods and parks throughout Europe.

Male and female peppered moths can be distinguished by their antennae. Those of female moths are slender and hairlike. Those of male moths are feathery and contain a battery of sensory organs that play a major part in locating a mate. Unmated female peppered moths emit a tell-tale odor from their bodies that attracts males. Carried along on air currents, this odor can be detected over a considerable distance by males flying in the same air currents, and these potential mates then home in on the female.

Twiglike caterpillars

The caterpillars of geometer moths are called loopers in Britain because of the way they arch their bodies when walking. In the United States they are called inchworms. The peppered moth caterpillar feeds on trees and bushes such as oak, elm, birch and bramble from July to September. It is just as inconspicuous on a tree as the adult moth because it resembles a twig. Green or brown with minute white dots, the caterpillar rests with its deeply notched head and body raised at an angle to a twig, hanging on with its hind legs. Between the two pairs of hind legs, or prolegs, are fleshy knobs that assist the camouflage by making it look as if the caterpillar is firmly attached to the twig by eliminating any shadow under the body. To complete the illusion, the caterpillar also has warts on its body that look like the joints of a twig. To pupate, the caterpillars burrow into the soil, and the adults emerge the following May.

Evolution in action

The study of the peppered moth has provided scientists with an example of evolution in progress, the gradual change in form as a result of natural selection. There are three color forms of the peppered moth. The typical form, *Biston betularia*, has white wings "peppered" with black specks that sometimes form faint black lines. A second form, *Biston betularia carbonaria*, is an intense uniform black color except for a white spot at the base of each forewing. It is often referred to as a melanic form. This term is derived from the name of the dark pigment, melanin, that gives the moth its color. The third color form, *Biston betularia insularia*, is an intermediate form that is black speckled with white.

Significant changes took place in the population of peppered moths in Britain and elsewhere during the 19th century. Before the Industrial Revolution the melanic form of the peppered

The dark (top) and normal (foreground) varieties of the peppered moth are simply different colored forms and can interbreed. By day they are prey for insect-eating birds; by night they are caught by bats while in flight.

When is a twig not a twig? A peppered moth caterpillar blends into its environment. It forms the left-hand side of the triangle in the center of the picture.

easily picked off by insect-eating birds. The black variety, meanwhile, which had been at a disadvantage, now flourished because birds failed to see it on lichen-free, soot-covered trees.

Over the years, scientists carried out experiments to demonstrate that industrial melanism was driven by such an alteration of circumstances, that one form was better fitted to survive than another under a given set of conditions. One of these experiments was carried out by H. B. Kettlewell. He placed dead peppered moths of the typical and black forms on tree trunks in an industrial area. Birds found 60 percent more of the typical moths. Furthermore, Kettlewell found that when peppered moths were kept overnight in a cage painted with black and white stripes, black forms usually settled on black stripes and typical forms on white stripes. Apparently the moths actively seek the right background.

moth, often simply called *carbonaria*, was extremely rare. Then, as towns and even countryside became coated with soot from homes and factories, melanic moths became more numerous and in some places the typical peppered moth completely disappeared. Biologists have long studied the peppered moth and have discovered that the dark forms are most frequently found in areas of Britain with high levels of air pollution. The normal form is common in rural areas such as the southwestern tip of England and the north of Scotland. Where smokeless zones have been introduced, the trend toward the dark forms is being reversed, the typical variety once more becoming common in areas that were formerly highly polluted.

Industrial melanism

The phenomenon of the change from the typical variety to the dark forms in areas of high pollution is called industrial melanism. Scientists thought that industrial melanism was probably the result of a change in the interaction between the moths and their predators. Peppered moths rest on lichen-covered trees by day, the typical form blending in with the lichen as a device to escape the attention of predators. With the advent of industry, soot in the air killed the lichens that provided camouflage for the typical form. As a result the moth showed up starkly against the dark background of the trees and was

PEPPERED MOTH

PHYLUM	**Arthropoda**
CLASS	**Insecta**
ORDER	**Lepidoptera**
FAMILY	**Geometridae**
GENUS AND SPECIES	***Biston betularia***

LENGTH
Wingspan: 1¾–2½ in. (4.5–6 cm)

DISTINCTIVE FEATURES
Adult (normal form): pale background with black specks giving peppered appearance; adult (intermediate form): black background peppered with pale specks; adult (dark form): almost entirely black with small white dot at base of each forewing; caterpillar (normal form): small size; twiglike appearance; green or brown in color

DIET
Caterpillar: leaves of oak trees (*Quercus*) and other deciduous trees and shrubs

BREEDING
Eggs laid early summer; autumn pupation

LIFE SPAN
More than 1 year

HABITAT
Temperate forest, including parks

DISTRIBUTION
Much of Europe

STATUS
Common

PERCH

THIS FISH, WHICH ORIGINALLY gave its name to the largest order of fish, the Perciformes, is the freshwater perch, *Perca fluviatilis*, of Europe. The name is derived from Greek and Latin, through the French, and it was known to the Romans as *perca*. There are many perchlike fish known today, so the main attention here is concentrated on the European perch. This plump-bodied fish is dark greenish with a yellowish tinge and dark bars on the flanks. The undersurface is silvery blue to yellowish and the anal and pelvic fins are reddish. The color varies, however, from one population to another, and in some localities the bars may be missing. The first dorsal fin is spiny, but both are well developed. There is a medium-sized anal fin, the pelvic fins are well forward and the tail is almost square-ended. A perch usually weighs about 1 pound (0.45 kg), although the record is about 10 pounds (4.5 kg) for a fish measuring 20 inches (50 cm).

The European perch is found in fresh water in much of Europe, western Asia and Siberia, and in slightly brackish waters around the Baltic Sea. Its counterpart in North America, east of the Rockies, is the yellow perch, *Perca flavescens*, which is golden with dark bars, a silver belly and orange anal and pelvic fins. A near relative in northern Europe is the ruffe, *Gymnocephalus cernua*, a somewhat smaller fish with a marbled pattern and lines of distinct dark spots on the fins. The walleye or pikeperch, *Stizostedion vitreum*, of North America, can attain three times the length of the European perch and has a blotched pattern. The name walleye stems from its prominent eyes, which have large, glassy pupils. Their white stare is a result of light reflected back through the pupil by crystalline matter in the retina. This enables the walleye to see well in darker waters. In North America there are also small darters, fast-moving, brilliantly colored fish, only a few inches long.

Lying in ambush

These effectively camouflaged predatory fish lurk among the stems of water plants, suddenly dashing out to seize their prey. The mouth is small but opens into a wide gape. The main sense is sight, but perch can hear and smell. There are two nostrils on each side of the head, one that takes water into the nasal pouch and the other at the rear that lets water out. Inside this pouch is a rosette of sensitive tissue.

Perch live in shoals in slow-flowing rivers and lakes. The smaller the fish, the larger the shoal, so at 3 years old they swim in small groups of a half dozen or fewer and later may even be solitary. In winter they retire to deeper water, as deep as 30 feet (9 m) in lakes, and remain quiescent. They can, if necessary, use the oxygen in the swim bladder for breathing.

Swallow fish head first

An adult perch eats smaller fish, which it usually seizes from behind with its sharp teeth, damaging the tail. It then swallows the fish head first. Fry (young) up to 1 month old feed on

The European perch, displaying its yellowish tinge and the dark bars on its flanks. There are many other species of perch, with 160 species in the family Percidae.

Bulrush stems provide a suitably dark and concealed location from which this European perch may launch an ambush on smaller fish.

PERCH

CLASS	Osteichthyes
ORDER	Perciformes
FAMILY	Percidae
GENUS	9 genera

SPECIES   159, including European perch, *Perca fluviatilis*; yellow perch, *P. flavescens*; ruffe, *Gymnocephalus cernua*; and walleye, *Stizostedion vitreum* (detailed below)

WEIGHT
Up to 24¼ lb. (11 kg)

LENGTH
Up to 67 in. (1.7 m)

DISTINCTIVE FEATURES
Long, slim body; prominent eyes with large pupils; brownish green or silver above to creamy white below; dark vertical stripes; prominent white margin on ventral lobe of tail fin; 2 dorsal fins, with large black spot at base of last three spines of first

DIET
Adult: insects and other fish; also crayfish, snails, frogs, mudpuppies and small mammals. Young: small invertebrates.

BREEDING
Spawns in small groups (1 larger female and 2 smaller males, or 2 females and up to 6 males); eggs and sperm released in single night; number of eggs: tens of thousands

LIFE SPAN
Up to 30 years, usually less

HABITAT
Lakes, backwaters, medium to large rivers; favors large, shallow, muddy expanses

DISTRIBUTION
Much of Canada and U.S.

STATUS
Common

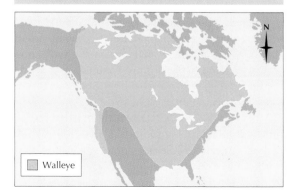

☐ Walleye

water fleas and other small plankton, after which they eat bottom-living invertebrates, such as midge and mayfly larvae, freshwater shrimps, and the occasional leech. During July, when the small perch are about 2 months old, there is a tendency toward cannibalism. Perch feeding only on fish grow faster than those forced to eat other food when fish are scarce.

Strings of eggs

Spawning takes place during April and May in the European perch, the fish shoaling according to body size. The eggs, like those of the American yellow perch, are laid in long strings that become entangled with water plants. The female glides over the water plants with her fins lowered, shedding the eggs, which are then fertilized by one or more males. A large female may lay 200,000 eggs. They hatch in about 18 days, or in only 8–10 days if the weather is warm. The transparent larvae are about 5 millimeters long. On hatching, each larva spirals to the surface to fill its air bladder. After this they hang for a while on water plants and then float at the surface. Perch mature at 3 years. The maximum recorded life span for the European perch is 10½ years; a walleye may live more than twice as long.

PÈRE DAVID'S DEER

THIS DEER HAS BEEN, to all intents and purposes, extinct in the wild for many years and is now known only from the descendants of a herd kept in the emperor's hunting park near Beijing (formerly Peking) in China. It was discovered in 1865 by the French missionary and naturalist Père Armand David.

In this deer the front prong of the antlers is forked but the hind prong is usually straight and slender. Unlike those of other deer, the antlers are sometimes shed twice a year, the summer antlers, measuring 2⅓–3 feet (70–90 cm) along the curve, being dropped in November. If they occur at all, the second, much smaller pair, are hard by January and dropped a few weeks later. The hinds (females) have no antlers. Père David's deer stands up to 4⅗ feet (1.4 m) at the shoulder and the coat is reddish tawny mixed with gray. There is a darker stripe across the shoulders and the underparts and a ring around the eye are white. There is a mane on the neck and throat, and the tail is tufted and longer than that of any other deer. The hooves are large and spreading.

Origin unknown

Even the Chinese did not know where Père David's deer originally came from, but it is thought to have inhabited the swampy plains and marshlands of northern China until cultivation of the land wiped it out, except for some kept by the emperors in their hunting parks. The species used to survive only in herds in Woburn Park, England, and various zoos throughout the world. However, in 1985 a herd was reintroduced near Beijing and this has since grown to 100 animals. Another group reintroduced near Shanghai now numbers around 120 deer.

Unlike most deer, Père David's deer is very fond of water. It swims well and will spend long periods standing in water, often up to its shoulders. Although predominantly a grazing animal, in summer Père David's deer supplements its grass diet with water plants.

Boxing stags

The rut begins in June in England, when the hinds group together in harems dominated by a stag (male). The master stag often engages in

fights with rival stags for the possession of the harem. He not only uses his antlers and teeth when fighting but also rises on his hind legs to box in the same way as the wapiti, *Cervus elaphus*. The master stag sires the early calves until he is driven off by another stag, and this goes on until the rut ends in August. The stags usually keep away from the hinds for about 2 months before and 2 months after the rut, but the sexes are together for the rest of the year. The one or two boldly spotted fawns are born in April and May after a gestation period of 288 days. The species' life span is about 20 years.

"The four unlikes"

Père David's deer has a quite unusual history. In 1865 Père David looked over the wall of the Chinese emperor's hunting park near Beijing and saw a herd of about 100 unusual-looking deer. The Chinese called the species *su-pu-hsiang*

After being wiped out by cultivation in China, Père David's deer were all but extinct in the wild until their recent reintroduction near Beijing and Shanghai.

Père David's deer are unusual among deer in that they swim well and spend long periods of time in water.

("the four unlikes") because it was credited with having the antlers of a stag, the neck of a camel, the hooves of a cow and the tail of a donkey. The Chinese believed the antlers had medicinal properties, and they also carved small works of art from them. No stranger, however, was allowed into the emperor's park and it was not until the following year that Père David managed, by bribing the Tartar guards, to obtain two skins to send back to Paris. Later several live deer were shipped across to Europe to various zoos.

In 1894 a disastrous flood in China breached the wall of the imperial hunting park and most of the herd of Père David's deer escaped into the surrounding countryside, where many were killed and eaten by the starving peasants. Most of the survivors were destroyed in 1900 during the Boxer Rebellion, and by 1911 only two remained alive in China. A few years later both of these were also dead.

The deer sent to the European zoos, meanwhile, did not breed very successfully and in 1900 the 11th Duke of Bedford decided that the only hope of saving the species was to collect all the survivors in one herd in Woburn Park. The famous herd was accordingly started with 18 deer. The herd flourished, and by 1939 it numbered about 250. During World War II it seemed far from sensible to keep all the living members of a species in one place. Accordingly, day-old calves from the herd at Woburn were sent to Whipsnade Zoo, and in time a small herd was also established there. Calves from Woburn and Whipsnade were subsequently sent to zoos in almost every part of the world, followed by the most recent reintroductions of these deer back into their native habitat in China.

PÈRE DAVID'S DEER

CLASS **Mammalia**

ORDER **Artiodactyla**

FAMILY **Cervidae**

GENUS AND SPECIES **Elaphurus davidianus**

WEIGHT
Male: up to 472 lb. (214 kg); female: up to 350 lb. (159 kg)

LENGTH
Head and body: 6–7⅓ ft. (1.8–2.2 m); shoulder height: 4–4⅗ ft. (1.2–1.4 m); tail: ¾–1⅕ ft. (22–36 cm)

DISTINCTIVE FEATURES
Red, tawny or gray-brown coat; darker stripe across shoulders; white ring around eye; longer tail than in other deer. Male: antlers of front prong forked, hind prong straight. Female: no antlers.

DIET
Grasses; also water plants in summer

BREEDING
Age at first breeding: 2 years; breeding season: June–August (England); number of young: 1 or 2; gestation period: about 288 days; breeding interval: 1–2 years

LIFE SPAN
Up to 23 years

HABITAT
Originally swampy plains and marshland; most animals now in captivity or semicaptivity in parkland or zoos

DISTRIBUTION
Native to northern China; large captive population in Woburn Park, southern England; two reintroduced herds near Beijing and Shanghai, China

STATUS
Critically endangered

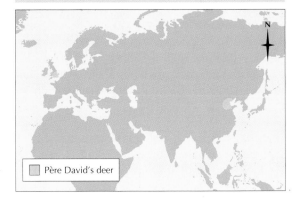

Père David's deer

PEREGRINE FALCON

THE PEREGRINE IS ONE OF the best-known falcons and a favorite among falconers. It is a large bird, 14–19 inches (36–48 cm) long, the female being larger than the male. Its upperparts are slate blue with darker barrings, while its underparts are white with black barring. Young peregrines are browner than adults and have streaked rather than barred underparts. The prominent black mustache stripes on each side of the face probably absorb light, reducing glare from the ground that would prevent the bird from seeing its prey clearly.

Peregrines are almost cosmopolitan, living on all continents except Antarctica and on many oceanic islands. They differ quite considerably throughout their range, both in size and coloring, and are consequently divided into several subspecies. The largest peregrines live in Arctic regions and the smallest species dwell in the deserts of North America and Arabia. Peregrines are found over most of Europe and much of Asia and Africa, including Madagascar. They also inhabit much of North America, including the far north, but are absent from the southern parts of the United States and most of Central America and South America, except at the southern tip of the continent. They are also found in Australia and on islands to the north of the country.

Falling numbers

Although they inhabit forests, open plains and moors, peregrines are most numerous in rocky areas or mountains, particularly sea cliffs. Each pair of peregrines defends a territory, the size of which depends on the abundance of food. On cliffs where seabirds nest there may be a pair of peregrines every 2 miles (3.2 km), but elsewhere territories may cover tens of square miles. When they are not feeding or caring for their young, peregrines roost on a favorite perch or circle their territory. Peregrines that breed in high latitudes migrate in winter. Some European peregrines, notably those of the subspecies *calidus*, cross the equator to countries such as South Africa. They follow well-defined routes, usually staying near to the coast but sometimes crossing open water. The young occasionally accompany the parents.

In common with many other birds of prey, peregrines have become less numerous over time. Their decline started with the spread of intensive agriculture and game preserves. During the first half of the 20th century, however, in Britain at least, the birds were recovering their numbers, though they suffered a setback in World War II when many were killed to protect

carrier pigeons that were being used as messengers. However, after about 1955 there was a dramatic decline in the numbers of peregrines in many parts of the world. This fall was due mainly to the effects of pesticides, which become concentrated in the eggs of birds of prey, rendering them infertile. Seeds tainted with tiny amounts of insecticides are eaten by small mammals and birds. Because they eat many seeds, large amounts of insecticides accumulate in these animals. Peregrines eat many small

Peregrines use their powerful claws to seize and kill their prey. Afterward they pluck the prey using the sharp, hooked bill before feeding.

The peregrine's large eyes enable the bird in flight to spot prey far below on the ground. Powerful eye muscles allow the peregrine to keep the prey in focus when diving for attack.

PEREGRINE FALCON

CLASS	**Aves**
ORDER	**Falconiformes**
FAMILY	**Falconidae**
GENUS AND SPECIES	***Falco peregrinus***

WEIGHT
Male: 1⅓–1½ lb. (582–750 g); female: 2–2⁹⁄₁₀ lb. (925–1,300 g)

LENGTH
Head to tail: 14–19 in. (36–48 cm); wingspan: 37½–43⅓ in. (95–110 cm)

DISTINCTIVE FEATURES
Broad-based, pointed wings; long, feathered legs; relatively short tail. Adult: black crown, mustache and nape; blue-gray upperparts; white underparts, finely barred with black. Juvenile: dark brown above; streaked below.

DIET
Mainly other birds; also small mammals; occasionally insects and frogs

BREEDING
Age at first breeding: 2 years; breeding season: varies according to location; number of eggs: 3 or 4; incubation period: 29–32 days; fledging period: 35–42 days; breeding interval: 1 year

LIFE SPAN
Up to 16 years

HABITAT
Extensive open terrain; nests on buildings, cliffs or treetops

DISTRIBUTION
Canada and parts of western U.S.; southern South America; much of Europe; Asia north of 50° N; southern Asia south into Australia; sub-Saharan Africa and Madagascar

STATUS
Scarce to common

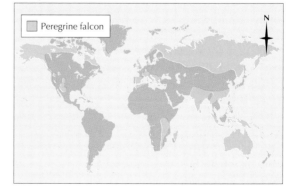

□ Peregrine falcon

birds, and in doing so build up a concentration of the insecticide, which damages embryos in the eggs. By the 1970s, the peregrine population was less than half its postwar average. However, DDT is no longer used in most countries and peregrine numbers are improving. From 62 pairs in 1962, breeding numbers rose to more than 1,000 pairs in the 1990s in Britain. There are similar increases in North America.

Stooping to conquer

Peregrines feed mainly on birds but also on mammals such as young hares, mice and voles and, occasionally, amphibians and insects. Pigeons are favorite prey, although smaller subspecies take a greater proportion of small birds. Peregrines often catch grouse on moors and take seabirds around cliffs.

Prey is caught after a swift dive, or stoop, performed by the peregrine with its wings nearly closed. The victim is either killed in the air by being struck with the peregrine's hind claws or is carried to the ground, and then to a special feeding place, where it is plucked before being eaten. The peregrine also sweeps birds from their perches, although the prey sometimes escapes by maneuvering violently or by keeping close to the ground. However, the peregrine also takes birds, especially poultry, directly from the ground.

Aerial exchanges

Peregrines mate for life and use the same nest site year after year. Sometimes the nest is a scrape in the ground; at other times the abandoned nest of a raven, gull or buzzard is used. The most popular nest sites are on cliffs and are often accessible only to expert climbers. Peregrines also occasionally nest on buildings.

At the start of the breeding season the male perches near the nest site. When a female appears he flies out and back, calling to attract her. During courtship the pair dive and swoop together or tumble in the air, calling frequently.

There are usually three or four eggs in each clutch, although there may be as many as six. They are laid at 2- to 3-day intervals and incubation starts before all the eggs are laid so they hatch at intervals after a period of about 1 month. The female does most of the incubation and is fed by the male, which brings food to her on the nest or passes it to her in the air or on a feeding perch. The female is very aggressive, attacking other large birds, humans and dogs that come near the nest. When the chicks first hatch, the female broods them, but after 2 weeks she covers them only at night. The male continues to bring food to the female, which, in turn, feeds it to the chicks, starting with the oldest. When the chicks are older, the male may give food to them directly. The young birds leave the nest when they are 5–6 weeks old and remain dependent on their parents for another 2 months, sometimes migrating with them.

Winged speedsters

Although it was probably not held in such high esteem as the gyrfalcon, *Falco rusticolus,* the peregrine was a royal bird in medieval falconry. Even in strong winds it flies under perfect control and is capable of remarkable accuracy when stooping, so it is not surprising that it was an extremely popular sporting bird, especially because it is easily tamed.

Peregrine falcons have been recorded flying at maximum speeds of about 170 miles per hour (272 km/h). However, flying at this speed poses more problems than simply those of supplying power and maintaining a streamlined shape. The bones, sinews and muscles must be strong enough to withstand the strains imposed at such velocities, especially during maneuvering and braking, while the birds' senses must remain extremely acute and the reactions must be quick. One problem that peregrines face is how to breathe air that is rushing past at a high speed. The nostrils are ridged, and within each nostril is a rod with two fins behind it. When air rushes past the nostril, the flow is broken up as it swirls past the rod and fins, so the birds need to exert little effort to suck it in. Other fast-flying birds of prey have similar structures, but slower species, such as vultures and sea eagles, lack them.

Despite their capacity for high speeds, peregrines do not usually exceed 60 miles per hour (96 km/h) in level flight. Some scientists have suggested that peregrines usually sacrifice speed in order to ensure greater accuracy.

Female peregrines brood their chicks for about 14 days after they hatch, feeding them with food brought by the males. The young fledge at 5–6 weeks of age.

PERIPATUS

Macroperipatus torquatus *from Trinidad in the Caribbean, is one of the largest of all the Onychophorans, reaching 5 inches (12.5 cm) in length.*

They live under stones, rotting wood, the bark of fallen trees and in similar damp places, being unable to withstand dessication. In a dry atmosphere they can lose a third of their weight in less than 4 hours, and will dry up twice as fast as an earthworm and 40 times as fast as a smooth-skinned caterpillar of a similar size. This is due to the unusual breathing system of the velvet worm; insects breathe through branching tubes (tracheae). Because the openings are few, there is little loss of water and there is also an efficient mechanism for closing the openings when necessary. Velvet worms have unbranched breathing tubes, so far more of them are needed to supply enough oxygen. Each tube has an opening, resulting in a rapid loss of water from the body when the surroundings are dry, which explains the limited distribution of these animals to the damper parts of the world.

AMONG THE MOST extraordinary of all animals living today, *Peripatus* and the other members of the order Onychophora are also among the most ancient, with fossil forms almost identical to those of today found in rocks dating back over half a billion years. Also known as velvet worms, they resemble both the soft-bodied annelid worms, such as the earthworm, and the hard-bodied arthropods, which include insects, spiders and crustaceans among their number.

The body of *Peripatus* is rather wormlike, tapering toward the hind end. In most species it is between ½–5 inches (1.25–12.5 cm) long, but can be extended or contracted, and it is sinuous in movement. Velvet worms are very variable in color, ranging from dark slate to reddish brown in the various species, and there is usually a dark stripe down the back. The skin is dry and velvety to the touch and there are up to 43 pairs of short, unjointed legs, each ending in a pair of hooks. There is a pair of flexible antennae on the head, with an eye at the base of each. The eyes are quite simple, although each has a lens. They are directed outward and upward and probably do little more than distinguish between light and darkness. The sensory hairs clothing the antennae and most of the body are specialized organs of touch and taste.

Life in damp places

Velvet worms depend on moist conditions, and live throughout the Tropics, as well as in damp forests throughout the Southern Hemisphere.

Sticky threads

Velvet worms mainly eat small insects such as termites, as well as other small animals such as pill bugs. In order to trap fast moving prey, the velvet worm shoots out a jet of milky white fluid from specialized nozzles, termed oral papillae, which are situated on either side of the mouth. On contact with the air, the fluid solidifies rapidly into sticky beaded threads of slime 3–12 inches (7.5–30 cm) long, which entangle the insects, allowing them to be caught and consumed. These sticky threads are also squirted over predators such as spiders, allowing the velvet worm to make its escape.

Casual lovemaking

Although some velvet worms reproduce through internal fertilization, the mating of *Peripatus* can only be described as casual. The male places capsules containing sperm on the female, apparently at random because he will even place them on her legs. White blood corpuscles in the blood of the female migrate to a point beneath a capsule and break through it by digesting the skin cells. At the same time the lower wall of the capsule breaks and the sperm enter the female's bloodstream, from where they eventually find their way to an ovary. There they force their way through the ovary wall. If an immature female receives sperm, the young egg cells feed on them and grow for a year before they are ready to be fertilized by a second mating. Except in a few species that lay eggs, the embryos develop in the uterus, taking in nourishment from the mother through its walls or from a specialized yolk-sac. In one South American species tissues form into a structure quite similar to the mammalian placenta, which passes nutrients to the growing embryos. In some species development takes 13 months and as young are born annually, there is one month in each year when a female is carrying two sets of embryos, one just beginning to develop and the other nearly ready to be born.

An ancient and mysterious group

Animals first appeared in water, and since that time there have been several independent invasions of the land. All velvet worms today are terrestrial, but several fossil forms have been found indicating that these animals were once found in marine environments. These fossils can be extremely ancient; for example, *Aysheaia* is from the middle Cambrian Burgess shale deposit, while *Xenusion* from China lived even earlier, at the very start of the Cambrian, making it one of the earliest animals currently known. Many of these early velvet worms had a variety of spikes and spines, in contrast to the smooth-bodied animals of today, but in many ways they were remarkably similar to their modern descendants. At some stage velvet worms made the transition from water to land, after which the marine branch of the group became extinct.

For many years, debate has raged about the affinities of the group. They appear to share many characteristics with both the annelid worms and the arthropods, and it was believed that velvet worms represented the direct descendants of the common ancestor of these two major groups. Recently, new fossils have come to light which suggest that velvet worms are closely related to arthropods, and may even be part of that group, while they are more distantly aligned with the annelids. For now at least, most authors place them in a separate phylum of their own, the Onychophora.

Many of the features of the velvet worm are simple and reflect the probable body plan of the earliest animals. Yet despite the fact that little has changed in terms of general morphology over a long period of time, the velvet worm survives and prospers in the modern world. Some characteristics, such as the presence in some species of a placenta, are, in fact, highly advanced when compared with other invertebrates.

PERIPATUS

PHYLUM **Onychophora**

FAMILY **Peripatidae and Peripatopsidae**

GENUS AND SPECIES **Around 75 species, including *Peripatoides novaezealandiae***

ALTERNATIVE NAME
Velvet worm

LENGTH
½–5 in. (1.25–12.5 cm)

DISTINCTIVE FEATURES
Long, thin body with velvety appearance; between 14 and 43 pairs of unjointed legs; head has nonretractible antennae

DIET
Small insects; many species feed on termites

BREEDING
Many species fertilize internally; male *Peripatus* deposits sperm package on body of female, which is ingested into bloodstream where sperm travel to ovaries; some species lay eggs, others bear live young

LIFE SPAN
Up to 6 years

HABITAT
Moist, humid environments such as under bark, stones and leaf litter

DISTRIBUTION
Equatorial regions; damp forests of Southern Hemisphere

STATUS
Locally common

This species of velvet worm, Peripatus novaezealandiae, *is from New Zealand, where it thrives in lush temperate forest.*

PERIWINKLE

The common periwinkle may be found on rocky shores worldwide. Also known as the edible periwinkle, it is widely caught for human consumption.

PERIWINKLES ARE ONE OF the many types of sea snails. The original Anglo-Saxon name for this mollusk was pinewinclan or winewinclan and was applied to the largest of the European species of the genus *Littorina*. The term periwinkle, or winkle, is now applied to several related species elsewhere in the English-speaking world, notably North America. Four periwinkle species are of particular zoological interest because they show how seashore animals are arranged in zones at different shore levels.

The original or common periwinkle, *L. littorea*, is the largest periwinkle species and has been collected by humans in large quantities for many years as food. Its coiled shell is up to 1½ inches (3.75 cm) high and is usually black, sometimes brown or red. The shell is ridged in young periwinkles and smooth in older individuals. The flat periwinkle, *L. littoralis*, with its rounded, flattened shell, never reaches more than ½ inch (1.25 cm) in diameter. Its color is variable, ranging from bright yellow or olive green to brown, black or striped. The rough periwinkle, *L. saxatilis*, has a ribbed shell, with a more pointed apex than the flat periwinkle. The shell is yellow or white and is ½ inch across and ⅝ inch (1.6 cm) high. The small periwinkle, *L. neritoides*, is conical and dark reddish brown in color. It grows to about ¼ inch (0.6 cm) high.

The periwinkle shell has a round mouth that is closed by a dark horny disc, or operculum, permanently attached to the animal's foot. This is popularly called the winkle head. The true head has a broad muzzle and two tapering tentacles with an eye at the base of each. The underside of the foot has a line down the middle, and a periwinkle crawls by waves of contraction passing forward along the two halves of the foot alternately. Breathing is by a single feathery gill in the mantle cavity. All periwinkles are vegetarian and, as in other snails, rasp their food to pieces with a horny tongue, or radula.

The rough and common periwinkles are found throughout the North Atlantic. The small periwinkle ranges from Scandinavia to the Mediterranean and Black Seas, while the flat periwinkle ranges from Iceland to the Mediterranean Sea.

A shore quartet

Four of the periwinkles, although closely related, survive in markedly different situations on the seashore, where they may be covered by the tide once every 11–12 hours or exposed to parching sun and wind with no more than an occasional splash of salty spray for many days on end.

The small periwinkle lives in crevices in the rocks up in the splash zone, the part of the shore above extreme high tide that is wetted by spray

PERIWINKLES

PHYLUM	**Mollusca**
CLASS	**Gastropoda**
ORDER	**Mesogastropoda**
FAMILY	**Littorinidae**
GENUS	***Littorina*; others**
SPECIES	**Many, including flat periwinkle, *Littorina littoralis*; common periwinkle, *L. littorea*; small periwinkle, *L. neritoides*; and rough periwinkle, *L. saxatilis***

ALTERNATIVE NAME
***L. littorea*: edible periwinkle**

LENGTH
Up to about 1½ in. (3.75 cm)

DISTINCTIVE FEATURES
Fairly thick, conical shell; older shells may feature tiny holes bored by small sponges

DIET
Microalgae, seaweed and lichen growing on rock surfaces

BREEDING
Sexes are separate. Breeding season: September–April (*L. neritoides*), all year (*L. saxatilis*), eggs laid March–October (*L. littoralis*), spring (*L. littorea*); number of young: up to 5,000 per season (*L. littorea*); hatching period: 6 days (*L. littorea*); larval period: 2–3 months (*L. littorea*).

LIFE SPAN
Usually 2–5 years

HABITAT
Intertidal zones on rocky shores; absent from areas exposed to many waves

DISTRIBUTION
North Atlantic, including West Coast of U.S. and coastlines of northwestern Europe and Scandinavia; Mediterranean; Black Sea

STATUS
Common or abundant

only at the time of high spring tides. It may be completely immersed in water when, particularly in winter, storms drive the tides higher than normal. It feeds on lichen, so it is more a land animal than an aquatic animal and can survive without water for about 30 days.

The rough periwinkle lives between mid-tide level and the base of the splash zone and is uncovered for long periods, twice a day. It is less tolerant of life in water than to being exposed to the air. Scientists who marked individual rough periwinkles and studied their behavior found they could survive out of water for long periods. The rough periwinkle feeds on seaweed.

The flat periwinkle lives farther down the shore than the two species discussed previously. It dwells among the larger types of seaweed, on which it feeds and under which it shelters to keep damp. It also feeds on sponges, especially the purse sponge, to some extent.

The common periwinkle lives from the level of low-water spring tides to high-water neap tides (tides of minimum range that occur at the first and third quarters of the moon), on bare rocks, among seaweed and stones or on sandy shores. There may be very dense aggregations of these periwinkles in shallow pools or in wet depressions. Its tracks over the sand are often seen when the tide is out. On hot days, at low tide, however, it tends to get in under seaweed or into damp crevices or to glue itself to rock surfaces. It gives out a slime from the foot as it withdraws into its shell, which glues the edge of the opening to the rock. At the same time it closes itself in with the operculum. Although tight enough to prevent it from drying up, its hold on the rock is so loose that a slight gust of wind may blow it off. The common periwinkle eats seaweed, particularly fragments broken away that are beginning to decompose.

Flat periwinkles hide in rock crevices and beneath stones during the day, emerging at night to forage for food.

The rough periwinkle is often found in close proximity to marine vegetation such as wrack (above), on which it feeds.

Different ways of breeding

The sexes are separate, females being slightly smaller than males. In all periwinkles there is a definite mating with internal fertilization. The small periwinkle spawns every two weeks from September to April, to coincide with the highest tides and the greatest amount of spray. The larvae spend a long time in the plankton and eventually settle far down the shore before making their way up to the splash zone on foot. In the rough periwinkle the young develop inside the female and are born as small replicas of the parents. Breeding takes place most of the year. The flat periwinkle lays its eggs in masses of jelly on seaweed from March to October and the young leave the jelly as tiny snails, without passing through a stage as swimming larvae. The common periwinkle spawns in spring, the larvae swimming about for 2–3 months before settling on the shore as tiny winkles smaller in size than a pinhead. The eggs are laid in capsules, with one to three eggs in a capsule. The capsules float and the larvae emerge from them after 6 days to swim in the plankton. Each female may lay up to 5,000 eggs in a season.

Food for many species

Many periwinkles are eaten by seabirds, such as gulls and shorebirds, while the tide is out. Oystercatchers, plovers, redshanks and other shorebirds feed on them, and when the tide is in bottom-feeding fish such as plaice eat them. However, they are protected to a large extent from animals that swallow them whole by being able to shut themselves tight in their shells, and thus pass through seabirds unharmed.

Humans have eaten periwinkles for many years. During the 19th century and in the early part of the 20th century, winkles were known as the pauper's food in England, and large quantities were eaten, after being cooked and extracted with a pin. This is the origin of the largely British expression "to winkle," meaning to extract. The history of periwinkle consumption is uncertain, but necklaces made of periwinkle shells have been found in prehistoric settings in caves.

From sea to land

Shore-living animals and plants tend to be arranged in zones, from high tidemark to low tidemark. In some the zonation tends to be obscured by local conditions, but one of the most clearly marked ranges is that of the periwinkles.

The arrangement of the four periwinkle species suggests how a marine mollusk might become land living. Three of the species feed on seaweed and one, the small periwinkle, feeds on lichen. Two species breathe by a gill in the mantle cavity that must be bathed with seawater. The other two, the small periwinkle and the rough periwinkle, have very small gills and the mantle cavity acts as a lung. It only needs the young to develop inside the mother and be born as small but fully formed snails, with no larval period, for a life on land to be attainable. This has already happened in the rough periwinkle.

PHALANGER

THE PHALANGERS make up one of the largest and most widespread families of Australian marsupials. There are some 14 species in the family Phalangeridae, including the ten cuscuses, genera *Spilocuscus* and *Phalanger*, and the three brush-tail possums, genus *Trichosurus*; both of these groups are dealt with elsewhere in this encyclopedia. This article is concerned mainly with the only other species of the Phalangeridae: the scaly-tailed possum, *Wyulda squamicaudata*.

The name phalanger is often thought to derive from the word *phalanges*, meaning toes and fingers. However, the term actually derives from the Greek word for web, and refers to the web of skin joining the small second and third toes of certain animals. The claws on these two toes are divided, and together they form a comb that the phalangers use to groom their fur. Kangaroos and wallabies also have small second and third toes on the hind feet, and these are joined by a web of skin too. However, kangaroos and wallabies have only four toes on their hind feet, whereas phalangers have five toes on each foot.

Treetop lifestyle

With rare exceptions phalangers live in the tops of trees, and rarely descend to ground level. The exceptions include the scaly-tailed possum, which sleeps in rock fissures during the day but climbs up into the trees at night to feed.

The phalangers' life in the trees recalls the lifestlye of the tree squirrels in other parts of the world, as does the manner in which many of them move through the trees. Phalangers go from branch to branch and from one tree to another, although in doing so they seem to glide rather than bound in the manner of the red and gray squirrels of the Northern Hemisphere. As a group, however, phalangers show a greater diversity in shape than that found in squirrels; some phalangers have bushy tails, while others have featherlike tails.

During the day phalangers rest in nests made of leaves and twigs. Most phalangers are wholly or nearly vegetarian, feeding on fruits and leaves, but some also eat insects.

Female phalangers have a pouch opening forward that contains either two or four teats. Most females carry one offspring at a time, although occasionally they may carry up to three. In general, the breeding and life history of phalangers seem to be similar to those of kangaroos and wallabies.

Infants at risk

Because they are nocturnal species and spend so much of their time in the tops of trees, most phalangers seem to have few predators. The greatest current danger to them is the felling of trees by humans in order to clear the ground. Perhaps one of their main natural hazards is revealed in studies concerning the social behavior of the brush tail possum. Each female has a territory of 2¾ acres (1.1 ha), and each male has one of 7½ acres (3 ha). Young males have difficulty finding a vacant territory and are persistently harried by the adults in occupation of the existing territories. Where this happens, there is a high infant mortality.

Possum with a scaly tail

The scaly-tailed possum was not discovered by science until 1917. It has a very restricted range, being limited to the coastal portions of west, northwest and north Kimberly in the state of Western Australia, and to a few islands nearby. Its distribution appears to be related to rainfall as it does not appear in areas that receive less than 36 inches (90 cm) of rainfall per year.

As its name suggests, the scaly-tailed possum has a largely naked tail, covered with scales. It has a very limited range in the savanna of northwestern Australia.

A common cuscus,
Phalanger orientalis.
Like all phalangers,
cuscuses have curved
and sharply pointed
foreclaws for climbing.
A web of skin joins the
second and third toes
of the hind feet.

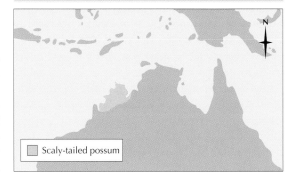

SCALY-TAILED POSSUM

CLASS	**Mammalia**
ORDER	**Diprotodontia**
FAMILY	**Phalangeridae**
GENUS AND SPECIES	***Wyulda squamicaudata***

WEIGHT
Up to 4½ lb. (2 kg)

LENGTH
**Head and body: up to 16 in. (40 cm);
tail: up to 12 in. (30 cm)**

DISTINCTIVE FEATURES
**Scaly, partly prehensile tail; large eyes;
small ears; pale or ashy gray upperparts;
white underparts; dark dorsal (back) stripe
runs from shoulders to rump**

DIET
**Fruits, blossoms and leaves of trees;
also some insects**

BREEDING
**Age at first breeding: 3 years; breeding
season: probably March–August; number
of young: 1; breeding interval: not known**

LIFE SPAN
Not known

HABITAT
**Regions of broken sandstone in hot
savanna woodland**

DISTRIBUTION
**Kimberly, in far north of the state of
Western Australia**

STATUS
At lower risk; extremely localized range

Scaly-tailed possum

The scaly-tailed possum has extremely large eyes, small ears, a sharp face and a coat of short, dense gray fur. Apart from its densely furred base, the rasplike tail is covered entirely in scales. This fact is alluded to in the animal's species name, comprising the Latin terms *squama*, meaning "scale," and *cauda*, meaning "tail." The tip of the tail is prehensile. The scaly-tailed possum appears to be a solitary animal, although fairly large numbers of the species are found in small areas of their favored habitat.

A wholly nocturnal animal, the scaly-tailed possum shelters by day among rocks, emerging at night to feed on the fruits, blossoms and leaves of trees such as *Eucalyptus* and *Terminalia*. It may forage during the day in the wet season, when there is heavy cloud cover. Although there is still scientific debate about the possible predators of the scaly-tailed possum, it is likely that they include wedge-tailed eagles, rufous and barking owls, carpet and olive pythons, dingoes and Gould's goannas. The possum's alarm call is a birdlike chittering.

In the few, small, remote localities where the scaly-tailed possum survives, the habitat is relatively pristine, and small populations seem to thrive. However, the species' range is so restricted that one misfortune, such as the planned bauxite mining in the region, could be disastrous.

PHALAROPE

PHALAROPES ARE SMALL shorebirds with needle-shaped bills and lobes on the toes, like those of coots, although not as well developed. Unusually for wading birds, the females are larger and more brightly colored than the males, although in winter both sexes have drab plumage.

The largest of the three species is Wilson's phalarope, *Steganopus tricolor*, which reaches about 10 inches (25 cm). The female's breeding plumage is generally slate gray above and white below, with reddish brown stripes on the back and throat. A broad black stripe runs through the eye and down the side of the neck. The male is mainly light brown above and white underneath.

The red phalarope, *Phalaropus fulicarius*, is slightly smaller, with a shorter bill. In summer the female has distinctive chestnut underparts, white patches over and behind the eyes and a yellow bill. The male is less colorful. In winter both sexes are gray above and white underneath. The name gray phalarope is widely used in Britain and aptly describes the winter plumage of these birds, which cross the Atlantic in the late fall. In North America the name red phalarope is appropriate for the birds' breeding plumage.

There are also alternative names for the third species, *P lobatus*: it is known as the red-necked phalarope in Britain and is sometimes called the northern phalarope in North America. It is the smallest phalarope, 6½–7½ inches (17–19 cm) long, and has a very slender bill. In winter it closely resembles the red phalarope, but in summer it is distinguishable by a black bill, slate gray head and upperparts, white throat and underparts and orange patches on the neck.

Phalaropes breed in northern parts of the world, migrating south in winter. The red phalarope breeds farther north than the other species, around the northern coasts of Alaska, Canada and Asia, as far as Spitzbergen. The distribution of the northern phalarope is similar but more to the south. Wilson's phalarope breeds inland in the central United States and Canada.

This female red phalarope is in full summer plumage. Phalaropes are unusual in that the females are brighter than the males.

The smallest phalarope is the red-necked or northern phalarope. This female, displaying the typical summer plumage, is swimming on its breeding pool. Red-necked and red phalaropes spend the winter on the open sea.

Shorebirds of the open sea

Although classified as shorebirds, phalaropes do not only wade in shallow water but also swim, floating high in the water because air is trapped in their very dense plumage. Wilson's phalarope is an inland bird and sometimes feeds on land, but the other two species nest near coasts, ponds, lakes or rivers, and are virtually seabirds outside the breeding season. They sleep afloat and rarely come to land during winter. Flocks of several hundred sometimes associate together on the sea. However, they do get blown by gales; they are swept across the Atlantic, for instance, and are driven inland, to be recorded eagerly by keen birdwatchers. Wilson's phalarope is rarely reported at sea. The salt gland, which is well developed in the other two species, is only rudimentary in Wilson's phalarope.

Stirring for food

Phalaropes feed by picking up small insects, crustaceans, worms and mollusks in their long bills. A few seeds are also eaten. Wilson's phalarope feeds more on land than the others and may be beneficial, taking large numbers of crane fly larvae and other pests. All three phalaropes have the strange habit of spinning around in the water like a top, at up to 60 revolutions a minute. This may just be the

RED PHALAROPE

CLASS	**Aves**
ORDER	**Charadriiformes**
FAMILY	**Scolopacidae**
GENUS AND SPECIES	***Phalaropus fulicarius***

ALTERNATIVE NAME
Gray phalarope (Britain only)

WEIGHT
1½–2½ oz. (40–75 g)

LENGTH
**Head to tail: 8–8¾ in. (20–22 cm);
wingspan: 8–8⅔ in. (20–22 cm)**

DISTINCTIVE FEATURES
**Needle-shaped bill; fleshy lobes on toes.
Breeding female: yellow bill; black crown;
white cheeks; chestnut underparts; black
and buff upperparts. Breeding male: duller;
lacks black crown. Nonbreeding (both sexes):
dark bill; gray upperparts; white underparts.**

DIET
Mainly invertebrates; some seeds

BREEDING
**Age at first breeding: 1 year; breeding
season: June–July; number of eggs: usually 4;
incubation period: 18–20 days; fledging
period: 16–18 days; breeding interval: 1 year**

LIFE SPAN
Not known

HABITAT
**Summer: marshy tundra near Arctic coasts;
winter: warm coastlines and offshore waters**

DISTRIBUTION
**Summer: Arctic, including Alaska, Canada,
Greenland, Spitzbergen and Siberia; winter:
Tropics and subtropics, especially waters off
western South America and West Africa**

STATUS
Common

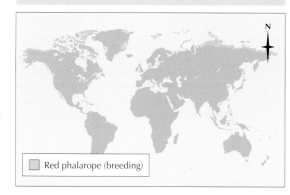

Red phalarope (breeding)

result of the birds continually turning to the rear to snap up food. However, the spinning may set up a small whirlpool that stirs up food from the bottom of shallow water or disturbs small animals, such as mosquito larvae, which can then be easily seen by their movements. Phalaropes also go up-end, like ducks.

The red phalarope has the habit of feeding on whale lice, settling on the backs of surfaced whales like a marine version of the oxpecker. During the Arctic summer it feeds on the backs of killer whales and belugas. In its winter feeding grounds in the Tropics, the red phalarope settles on sperm whales and on masses of floating seaweed to feed on small animals living there.

Topsy-turvy courting
Like button quails and painted snipe, the breeding habits of phalaropes are unusual in that the brightly colored and aggressive females display to each other and establish territories from which they court the attention of the less brilliant males. Sometimes several females may pursue one male.

Phalaropes breed in colonies, sometimes quite large, with the nests well spread out. Both sexes build the nest, which is a hollow lined with grass. The four, rarely three, eggs are incubated by the male alone for about 3 weeks. The female remains near the nest and sometimes helps in tending the chicks.

Chemical breeding control
The male and female differ in both form and behavior. In phalaropes these differences are reversed in that the courtship and incubation roles have changed without any alteration in the basic roles of the sexes in mating and egg-laying. The two are usually intimately linked; a castrated male chicken changes from an aggressive cockerel to a timid capon because hormones secreted by the testes and ovaries control the behavior associated with breeding.

In phalaropes the hormones have been reversed. Each sex usually has both male and female hormones, but females have very little male hormone and vice versa. However, female phalarope ovaries secrete large amounts of androgen, the male hormone. This hormone is responsible for their bright breeding plumage and aggressiveness. Male phalaropes incubate the eggs and develop brood patches, or areas of bare skin on the breast that keep the eggs warm. This behavior is controlled by the usually female hormone prolactin. Males have three and a half times more of this hormone than females.

Wilson's phalarope is a strictly inland species, unlike the other two phalaropes, which winter out at sea. All phalaropes feed on a range of prey, but they favor midges and gnats.

PHEASANT

THE NAME PHEASANT COMES from the Greek, meaning "the bird from the Phasis River" in the country formerly known as Colchis, to the east of the Black Sea. There are two species of true pheasants, genus *Phasianus*, but the name has been extended to include monal pheasants, eared pheasants, gallo pheasants, long-tailed pheasants, ruffed pheasants, peacock pheasants and others. These are all large, attractive birds with long tails. In general, the males have brighter plumage than the females.

The two species of true pheasants are the green pheasant, *Phasianus versicolor*, of Japan, and the ring-necked pheasant, *P. colchicus*, found across much of Europe, in parts of Central Asia and in eastern China. The latter has also been introduced into parts of North America and New Zealand. The male ring-necked pheasant has reddish brown plumage marked with orange, black and red, and a long tail barred with black, but the general effect is of burnished copper. Its head and neck are dark green and some males have a white necklace, giving this species its common name. It has red wattles surround the eyes and there is a pair of earlike tufts on the head. The female has far duller plumage, being entirely dun brown in color. The cock pheasant is 2½–3 feet (75–90 cm) long, of which 1–1½ feet (35–45 cm) is tail, while the hen is 1¾–2 feet (53–60 cm) long, of which 1 foot (30 cm) is tail.

The so-called game pheasant, found in many countries where the ring-necked pheasant did not naturally occur, is the product of introduced stock. Birds have been bred especially for their meat, and cross-breeding has produced many variants on the original wild birds' plumage.

Other pheasants

The three species of monal pheasants, genus *Lophophorus*, of the Himalayan region are thickset birds with short, square tails. The males are metallic green, blue, purple and coppery red above and velvety black below. The hens are streaked brown. The 11 species in the genus *Lophura* include the firebacks and silver pheasants. They range from the Himalayas south to Malaysia and have arched tails and velvety wattles on the face. The males are black, blue, purple or white above, with black underparts. In the three species of eared pheasants, genus *Crossoptilon*, of China the sexes are alike. Their plumage is hairlike, the tail is large and there are

The ring-necked pheasant (male, below), along with its 20 or 30 subspecies, is reared as a game bird across Europe and has now been introduced to North America.

RING-NECKED PHEASANT

CLASS	**Aves**
ORDER	**Galliformes**
FAMILY	**Phasianidae**
GENUS AND SPECIES	***Phasianus colchicus***

WEIGHT
**Male: 2⅕–3¾ lb. (1–1.7 kg);
female: 1⅗–2⅔ lb. (0.75–1.2 kg)**

LENGTH
**Head to tail: male, 2½–3 ft. (75–90 cm);
female, 1¾–2 ft. (53–60 cm); wingspan:
2½–3 ft. (75–90 cm)**

DISTINCTIVE FEATURES
**Chickenlike bill; strong legs; very long tail
with dark bars. Male: dark green head with
1 pair of small tufts; bright red wattles
around eyes; orange, red and black-marked
plumage; thin white necklace in some
individuals. Female: dun brown all over.**

DIET
**Grain, seeds, berries, fallen fruits and green
shoots; occasionally invertebrates, lizards,
young snakes and small mammals**

BREEDING
**Age at first breeding: 1 year; breeding
season: February–June; number of eggs: 7 to
14; incubation period: 23–28 days; fledging
period: 12 days; breeding interval: 1 year**

LIFE SPAN
Up to 8 years

HABITAT
**Flat farmland; hilly areas close to
cultivation; overgrown edges of rivers**

DISTRIBUTION
**Europe, east through Central Asia to eastern
China and southeastern Siberia; introduced
into North America and New Zealand**

STATUS
Locally common

Ring-necked pheasant

*Silver and fireback
pheasants, such as
this Siamese fireback,
Lophura diardi, are
distinctive for their
arched tails and the
velvety wattles on the
face. In the species of
this genus the sexes
often live separately
for most of the year.*

velvety wattles on the face and long white
feathers on the sides of the head. The five species
of long-tailed pheasants, genus *Syrmaticus*, range
from Japan and eastern China south to Myanmar
(Burma) and Thailand. The cock is yellow,
coppery red and dark blue marked with black,
gray and white. It has red wattles and a long
barred tail. The relatively drab hen is brown
marked with black and buff.

The ruffed pheasants include the magnificent
golden pheasant, *Chrysolophus pictus,* and Lady
Amherst's pheasant, *C. amherstiae*, from the
mountains of central and western China. The
cocks have a crest and a ruff, and the plumage is
yellow and red in the golden pheasant, green
and blue with white in Lady Amherst's pheasant.
The seven species of peacock pheasants, genus
Afropavo, of the forests from eastern India to
Borneo, are small with long tails, gray and brown
plumage, marked with metallic green and brown
eyespots and have two spurs on each leg.

Unwilling but powerful fliers

Like all pheasants, ring-necked pheasants are
ground birds, feeding and nesting on the ground
and roosting in trees. Their wings are short and
rounded, and although they do not fly far, their
flight is strong and fast. If alarmed, they run
rapidly over the ground or take off by flying

A male Himalayan monal pheasant, Lophophorus impeyanus, *sunning himself. The Himalayas and Southeast Asia are particularly rich in pheasants.*

almost straight up with whirring wingbeats. Although flight looks labored, pheasants can reach fast speeds of 50 miles per hour (80 km/h) or more over short distances. Their legs and toes are strong and spurred, and the claws are used to scratch the ground for food. Ring-necked pheasants feed on insects, seeds, berries, fallen fruits and leaves, worms, slugs, snails, lizards, even small rodents and young snakes. Roots, bulbs and tubers are also unearthed and eaten.

Large clutches
Breeding takes place in the ring-necked pheasant from February to June. Cock pheasants are usually polygamous, mating with several females, although there are reports of monogamy when hens are scarce. During courtship the cock displays to the hen by blowing up his wattles, puffing up his feathers and parading in front or to the side of her, with the wing nearest her drooping and his tail curved toward her.

The hen makes a nest by scraping a shallow depression in the ground and lining it with leaves and grasses, usually under coarse grass, bracken or brambles. She lays 7 to 14 olive brown eggs and these hatch after 23–28 days.

Incubation is usually by the hen alone, with the male helping only exceptionally. The young are able to fly after about 12 days.

Distracting predators
Once called injury feigning, the trick used by a hen pheasant when the nest is in danger is now known as distraction display. If a predator such as a fox comes near when there are eggs or chicks in the nest, the hen runs from the nest, trailing a wing, as if injured. The fox follows what looks like a disabled parent expecting an easy meal. The chicks, meanwhile, slip off silently and unobtrusively into the cover of adjacent vegetation and stay still, blending with the background color. The hen leads the predator away for some distance before taking wing and, flying in a wide curve, returning to a point near the nest and disappearing into cover until all is safe.

In their native habitat pheasants are killed by the usual predators, and their eggs and chicks are also taken. Domestic cats are known to wait at dawn for pheasants to come down from their roosts, and this is probably the pattern of hunting used by natural ground predators in the pheasants' native homes.

PICULET

PICULETS ARE SMALL woodpeckers that reach 3–5 inches (7.5–12.5 cm) in length. In appearance and habits they are more like nuthatches than typical woodpeckers. One important difference is that the tail feathers are not stiff and pointed, so the tail does not have the tattered appearance found in woodpeckers. One toe, however, is turned back, as in the true woodpeckers. There are 30 species of piculets. Three live in Southeast Asia, one in western and central Africa and the rest in tropical Central and South America and the Caribbean. They are generally gray or green above with spotted or streaked underparts.

The olivaceous piculet, *Picumnus olivaceus*, has olive-green upperparts and an olive wash to its underparts. Its crown is black with white dots; the male's forecrown is marked with fine red streaks. The Antillean piculet, *Nesoctites micromegas*, native to Gonave Island and Hispaniola, has greenish upperparts and yellow-white underparts streaked with black. Both sexes have a yellow crown, but that of the male has a red center. The rufous piculet, *Sasia abnormis*, of Peninsular Malaysia, is olive green above and rufous below. The male's forehead is yellow, while the female's is rufous.

Piculets hunt in tropical forests by moving among the fine twigs in the tree canopy, agilely working their way through them and hanging at all angles, as tits and titmice do. They go about in pairs or small family groups. Sometimes piculets work on trunks in the manner of nuthatches, but because they lack the stiff tail feathers of typical woodpeckers, they do not press their tails against the trunk as a support while they are hammering.

Hammering for insects

The bill of a piculet is the same shape as that of a true woodpecker, although it is smaller relative to body size. Like a woodpecker, the piculet feeds by hammering at wood or searching bark for insects to eat. It can drill holes only in the softest wood but hammers more vigorously than tits or nuthatches, which also search for insects that are hidden in wood. The olivaceous piculet is fond of eating ants. Insect larvae and pupae are also taken.

Nest drill

Despite their comparatively small bills, piculets carve out their own nest chambers. The rufous piculet is often found in bamboo jungles. It makes its nest chamber simply by boring a hole in a bamboo stem. The olivaceous piculet drills out its chamber, choosing trees that have decayed to almost the softness of balsa wood. The nest is built near the ground, not more than about 5 feet (1.5 m) up, and is unlined. Both sexes drill out the nest, sometimes working together, exchanging trilling calls as they work. One pair kept under observation took about a week to complete the nest.

Afterward both birds roost in the nest at night and by day they take turns to incubate the glossy white eggs. The chicks are naked and blind when they hatch. The parents seem to feed them mainly on the larvae and pupae of ants.

Piculets are tiny birds that belong to the woodpecker family. The speckled piculet (above) is one of three species native to Southeast Asia.

I.LYCETT

The olivaceous piculet is the only species of piculet found on the American mainland north of Panama.

SPECKLED PICULET

CLASS	**Aves**
ORDER	**Piciformes**
FAMILY	**Picidae**
GENUS AND SPECIES	***Picumnus innominatus***

WEIGHT
⅓–½ oz. (9–13 g)

LENGTH
Head to tail: 4 in. (10 cm)

DISTINCTIVE FEATURES
Tiny size; short, pointed bill; one toe turned backward on each foot; uniformly olive green upperparts; pale underparts with bold black spotting; thick black stripe behind eye; red forehead (male only)

DIET
Insects, particularly ants, and their larvae

BREEDING
Age at first breeding: 1 year; breeding season: January–May; number of eggs: 3 or 4; incubation period: about 11 days; fledging period: about 11 days; breeding interval: 1 year

LIFE SPAN
Not known

HABITAT
Deciduous forest; evergreen montane forest; open secondary forest (forest that has regrown after being felled), especially in areas with plenty of bamboo; found up to altitude of 9,900 ft. (3,000 m)

DISTRIBUTION
Himalayan foothills and Western Ghats (in India) east to central and southern China, south to Myanmar (Burma), Thailand, Vietnam, Malaysia and Sumatra

STATUS
Common or fairly common

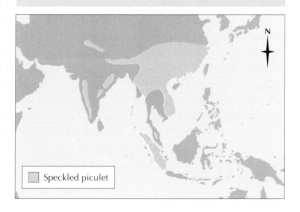

Speckled piculet

days old but return to the parents at night for another 1–3 months. Olivaceous piculets often have to make new nests or roosting holes as the rotten trees they choose are likely to fall down.

Long tongues

One of the distinguishing features of the woodpeckers, including the wrynecks and flickers (both of which are described elsewhere in this encyclopedia) as well as the piculets, is their extraordinarily long tongues. The tongues are used to wipe up or spear insects, often from narrow crevices or holes. The tongue is supported on a long, narrow bone that divides into two horns. This structure is common to all bird species but in the woodpecker family the horns are exceptionally long. They pass from the base of the lower mandible (bill-half) under the skull, around the back and over the top. In the piculets and in some of the flickers and woodpeckers, the horns stop at the base of the upper mandible, but in other flickers, and in the wrynecks and woodpeckers, they continue into one side of the upper mandible or curl under the right eye. A muscle runs from near the tip of the lower mandible and along the length of the horns. When it contracts, the horns are pulled forward and pressed against the skull, forcing the tongue out.

PIDDOCK

PIDDOCKS ARE MARINE, bivalve mollusks that bore into stone, wood, peat and sand with a rotary action of the shell. A piddock's shell is broader at one end than the other. There are prominent short spines where the concentric ridges and radiating lines on the shell meet. These are especially pronounced at the broader end where they form rasping files. Some piddock species bore only to the depth of the shell, while others bore much deeper. When in the burrow, a pair of long siphons reach up toward the mollusk's mouth. If disturbed, the piddock discharges a jet of water through these siphons. The siphons are united right up to the tip and in deep borers are often partly covered and protected with the horny material that covers the shell. One siphon draws in a current of water, bringing oxygen and food particles, while the other gives out a waste-carrying current.

Most species occur at the low intertidal or shallow subtidal zone, rarely at depths below 165 feet (50 m), but there are a few deep-sea piddocks.

The name piddock came into use in the early 18th century but it is not widely used outside Britain. These mollusks are therefore known, in North America, for example, as rock borers, rock-boring clams or pholas, from the scientific name of the best-known of them. Wherever they occur, piddocks are very similar to the European species in appearance and habits.

Scraping action

Besides having a filelike end, the shell must be made to scrape. This is made possible by an unusual arrangement in the hinge. In most piddocks the ligament found in other bivalves has been lost and the hinge teeth have been reduced to a double ball joint. This joint allows the two valves of the shell to rock on each other in a seesaw movement, by alternate contractions of two muscles, one in front of and one behind the hinge. In other bivalves, these muscles run from valve to valve and contract in unison to close the shell. In the piddock, the muscle nearest the front is the larger of the two, and part of it is spread outside the shell and joins the valves above the hinge. This exposed muscle is often protected by extra plates of shell.

At the front of the shell where the foot emerges, there is in most species a permanent gape. Apart from this region, the flaps of mantle tissue that lay down and line the shell are joined up, so the gills are not visible. This almost complete enclosure of the gills within the mantle cavity is common in boring and burrowing bivalves.

There are many species of piddocks. These include the common piddock, *Pholas dactylus*, the largest species at up to 6 inches (15 cm) long. It ranges from southern Britain to Morocco, the Mediterranean and the Black Sea. The little piddock, *Barnea parva*, 2½ inches (6.4 cm) long, makes horizontal rather than vertical burrows. The great piddock, *Zirfaea crispata*, 3½ inches (9 cm) long, is found around the northern coasts of Europe as well as the British Isles. The paper piddock, *Pholadidea loscombiana*, is also known as the American rock borer. It is peculiar in that when it has finished burrowing and is at most 1½ inches (3.8 cm) long, the gape in the shell through which the foot is pushed out becomes closed off by extensions of the shell, and the shell becomes trumpet-shaped at the base of the siphons.

Rotary borers

Except in the paper piddock, boring generally continues throughout the piddock's life. As the animal grows, the inner end of the tunnel becomes wider than the older part, bored when the piddock was younger. The piddock cannot leave its burrow, and any piddock removed from its burrow cannot make another one. Except in the white piddock, *Barnea candida*, which is more suited to boring in softer mate-

A common piddock, the largest of the piddock species. This rear view photo clearly shows the mollusk's gaping valves and foot.

PIDDOCKS

PHYLUM	**Mollusca**
CLASS	**Bivalvia**
ORDER	**Eulamellibranchia**
FAMILY	**Pholadidae**

GENUS AND SPECIES **Common piddock,** *Pholas dactylus*; paper piddock, *Pholadidea loscombiana*; little piddock, *Barnea parva*; great piddock, *Zirfaea cristata*; others

ALTERNATIVE NAMES
Rock borer; rock-boring clam; phola

LENGTH
Up to 6 in. (15 cm); most species less than 3½ in. (9 cm)

DISTINCTIVE FEATURES
Distinctive sculpture on shell with short spines where concentric ridges and radiating lines meet; spines especially prominent at front, where they form rasping files; shell very thin and brittle; distinctive shape with prominent gape in shell

DIET
Tiny planktonic particles

BREEDING
Eggs and sperm released into water where fertilization takes place; larvae spend weeks or months in plankton before settling

LIFE SPAN
Probably 5–10 years or more

HABITAT
Mainly low intertidal and shallow subtidal zones; some species in deep sea; bore into compacted sand, mud, clay, peat, wood and soft stones such as chalk and limestone

DISTRIBUTION
Almost worldwide

STATUS
Abundant

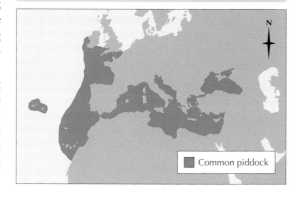

Some piddocks, such as this common piddock, will bore downward to depths several times the length of the shell.

rials, the foot ends in a broad sucking disc, and as the valves open the shell is rotated by the foot muscles, alternately one way and the other to make a circular hole.

Feeding mechanism

Piddocks feed in the same way as other bivalves. Minute planktonic particles are sucked in through one siphon and strained from the water by the gills, caught in mucus and passed to the mouth. The water currents and the highly organized streaming of the food particles over the gills, across the sorting areas of the lips (labial palps) to the mouth and even inside the stomach, are all the work of the cilia (minute, vibrating hairs). Sand and other rejected particles are wafted by cilia into the current in the outgoing siphon, where they are joined by feces and materials from the excretory organs.

The sexes are separate. During the breeding season, eggs and sperm are propelled by the current in the outgoing siphon into the open sea, where fertilization takes place. On hatching, the piddock larvae spend weeks or months in the plankton. Eventually they settle on a suitable surface and begin boring.

PIG-TAILED MACAQUE

PIG-TAILS ARE BIG MONKEYS, the largest of the macaques, but distinctive enough to be treated separately. The male weighs 14 pounds (6.4 kg), the female 6 pounds (2.7 kg). The coat is buff or brown, and the distinguishing feature of the pig-tail is the whirl or parting in the middle of the crown of the head, from which radiates a cap of short, dark brown or blackish hairs. The hair around the cheeks is light-colored and outwardly directed. The face has a long, pale brown muzzle and paler eyelids. The short, thinly haired tail, only one-third the length of the body, is carried arched over the back with the tip resting lightly on the rump.

Lion-tailed relative

The lion-tailed macaque, *Macaca silenus*, is much smaller, the males weighing only 3 pounds (1.4 kg), and in it the distinguishing features of the pig-tail are almost caricatured. Its fur is black and there is the same short-haired cap on the crown as in the pig-tail. The cheek hairs form a long, gray ruff, and the whitish eyelids are conspicuous against the black face. The tail is longer than that of the pig-tail, at about half the body length, and may be carried arched forward or backward. It, too, is short-haired but, as if to emphasize the thinness of the tail hair, there is a tuft at the end. "Lion-headed" would be as apt a description of this monkey.

Pig-tails live in Myanmar (Burma) and Vietnam south to Peninsular Malaysia, Sumatra and Borneo. They have been introduced into the Andaman Islands. Lion-tails are restricted to a small area in the Western Ghats of southwestern India, at altitudes of 2,500–4,300 feet (750–1,300 m), where the trees grow to 70 feet (21 m) or more. They are rare and still decreasing, being found recently in only four hill ranges: the Nilgiris, Anaimalis and Cardamoms and the range around Lake Periyar.

Whooping troops

Pig-tails and lion-tails both live in tropical rain forest, spending more time in the trees than most macaques. They live in large troops: one researcher's two lion-tail troops had 16 and 22 members, respectively, while in Peninsular Malaysia another zoologist found two pig-tail troops of 30 and 47, respectively. The troops have overlapping home ranges, which vary in diameter.

Each troop contains two or more fully adult males, one of which is dominant and leads the troop while another, or a young male, brings up the rear. Some solitary males live in and around the troop areas. When two troops approach one another to feed on fruit trees that are in the overlap area of their home ranges, the adult males whoop loudly at each other. The smaller troop usually moves away after a few minutes. Fighting has never been seen between troops. A large troop may break into two for some hours before rejoining.

Dissecting nuts

Like other macaques, pig-tails and lion-tails have a varied diet. They eat mainly fruit and leaves but also nuts, flowers, buds, pith, grubs, lizards and birds and birds' eggs. They pick nuts apart, meticulously peeling them with both teeth and fingers before eating them.

Breeding

Breeding takes place all year, with a peak from October to February. A mating pair is often found far from its troop. A single young is born after 162–186 days; it has brown hair and flesh-colored skin. After 1 month the skin becomes pale brown and then gradually turns black. The

A pig-tailed macaque forages for food on the forest floor. Although they feed mainly on vegetation, pig-tails also eat small vertebrates and take eggs from birds' nests.

PIG-TAILED MACAQUE

CLASS	**Mammalia**
ORDER	**Primates**
FAMILY	**Cercopithecidae**
GENUS AND SPECIES	***Macaca nemestrina***

WEIGHT
Up to 40 lb. (18 kg)

LENGTH
**Head and body: up to 30 in. (75 cm);
shoulder height: 28–39 in. (0.7–1 m);
tail: up to 15 in. (38 cm)**

DISTINCTIVE FEATURES
**Buff, brown or blackish fur, turning silver
with age; cap of short, dark brown or
blackish hairs radiating from parting on
crown of head; slight, light-colored ruff on
face; small, thinly haired, twisted tail**

DIET
**Mainly fruits and leaves; also nuts, flowers,
buds, pith, grubs, bird eggs and small
vertebrates such as lizards**

BREEDING
**Age at first breeding: 2–3 years (male),
3–4 years (female); breeding season: all
year, peaks October–February; number of
young: usually 1; gestation period: 162–186
days; breeding interval: about 1 year**

LIFE SPAN
More than 25 years in captivity

HABITAT
Rain forest and village edges

DISTRIBUTION
**Myanmar (Burma) and Vietnam south to
Peninsular Malaysia, Sumatra and Borneo;
introduced into Andaman Islands**

STATUS
Vulnerable; population: 100,000

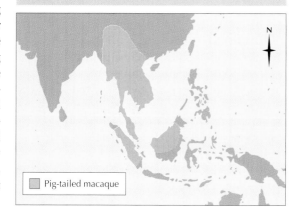

Pig-tailed macaque

*Tropical rain forests
are the preferred
habitat for pig-tails,
where they live in large
troops, overseen by
two or more adult
males. One of these
males is dominant and
heads the troop.*

hair of the lion-tail turns completely black except
for the face ruff. Only the cap on the head turns
blackish in the pig-tail. Baby pig-tails are weaned
at 8 months. One pig-tail is known to have lived
for more than 26 years in Milwaukee zoo.

Survival by cross-breeding

Tigers and leopards may kill pig-tails and lion-
tails but humans are their only serious enemy.
Lion-tails have been heavily overhunted, and it
may be that fewer than 1,000 of these monkeys
still exist. Pig-tails are more resilient, but even
they have suffered from hunting. In heavily cultivated areas the small crab-eating macaque, *M.
fascicularis*, often manages to survive, but the
larger and more conspicuous pig-tail gets shot or
otherwise squeezed out.

Irwin Bernstein, studying monkeys in
Peninsular Malaysia, came across an interesting
situation in a highly cultivated area. A crab-eater
troop survived in a forest isolated among the
fields. The troop thrived on crop-raiding; among
the crab-eaters were two monkeys that were
obviously hybrids between crab-eaters and pig-
tails. Bernstein surmised the pig-tails that had
inhabited the forest had been almost extermi-
nated, and that the one or two survivors had
joined a troop of crab-eaters, with which they
freely interbred. Finally, they too were killed,
and only these hybrids remained to show that
pig-tails had ever been there at all.

PIKA

PIKAS ARE SMALL MAMMALS related to the rabbits and hares. They are known by a variety of names, including mouse-hares, rock rabbits, rock conies, calling hares, piping hares and whistling hares. There are two species in North America and 24 species in Asia; the largest is 1 foot (30 cm) long, and the smallest is less than half this. Pikas look like rabbits or hares with short rounded ears and no tail. Each foot has five toes, and the soles of the feet are hairy, enabling them to run easily over smooth rock surfaces. The fur of pikas is usually grayish brown above and lighter on the underparts; it is reddish in one species. In general, the coat is lighter colored in dry areas and darker in more humid regions. Some species have two molts a year, giving a summer coat that is reddish or yellowish and a gray winter coat.

The North American pikas live in and around the Rockies, in the Sierra Nevada, Utah and New Mexico, and in southeastern Alaska and the Yukon in the north. In northern Asia pikas range from the Volga River and Ural Mountains to Korea and the Japanese island of Hokkaido, and in southern Asia from Iran to Nepal.

From lowlands to Mount Everest

One species of pika lives on Mount Everest up to 17,500 feet (5,250 m), the highest altitude at which any mammal has been found. Pikas live in a variety of habitats: on plains, in deserts, in forests and on rocky mountainsides.

One of the most noticeable features of pikas is their voice, which is usually a whistle or a sharp bark, *ca-ak*, repeated many times. Both calls are remarkably ventriloquial, the body being jerked forward and upward at each cry. Pikas rely for safety on remaining hidden, dropping into a crevice and there lying still. Among rocky screes they use the crevices and cavities as shelters, while on the plains they burrow.

Making hay while the sun shines

Pikas usually live in places where the winters are cold, but they do not hibernate. Instead they have the remarkable habit of cutting vegetation with their chisel-like teeth, drying it in the sun and storing it for winter fodder. A pika may travel several hundred feet from home to cut herbs and grasses, carrying these in its mouth to a chosen spot to dry and adding a fresh layer each day. Some pikas climb into the lower branches of young trees to take young green shoots. In winter bark is sometimes eaten as the pikas tunnel under snow, but the main food even then is the dry

fodder. This fodder is stored under an over-hanging shelf of rock or under a fallen tree, a single store holding a bushel of hay. Pikas feed in the early morning and late afternoon. Midday is spent basking. During the day the droppings are small, green and dry. At night they are black and wrapped in a jellylike layer that keeps them soft and wet. These droppings are swallowed again and kept in the stomach to be mixed with fresh food and redigested, a habit first noticed in rabbits. It has been found that if rabbits are prevented from eating their soft night droppings they will die in about 3 weeks. The droppings are their only source of certain essential vitamins, formed by the activity of bacteria breaking down partly digested plant material in the droppings.

The miners of the Yukon and elsewhere in the western half of North America called the pika the starved rat. Although strictly vegetarian, pikas must nevertheless be well fed and are far from deserving this nickname. Animals in the

In summer, the pika collects grass stems and other vegetation, leaves them to dry in the sun and then piles them into small haystacks. These dried food stores are eaten during the winter months.

A distinctive whistle or bark is a characteristic feature of the pika. The animal is something of a ventriloquist, since neither call seems to emanate from the creature itself.

North Temperate Zone can stand up to cold so long as they are well fed. It is not the hard winters that kill, but food shortages caused by freezing conditions. Pikas can keep going even when the ground is covered with snow because they have their food stores. They even sun themselves on rocks in temperatures of 0° F (-17° C).

Small, naked babies

The breeding season is often between May and September, when each female has two or three litters. The gestation period is 30 days and there are usually two to five babies in a litter. Each young is born naked and helpless and is put in a nest of dried grass. They weigh 1 ounce (28 g) at birth; adults weigh up to about 14 ounces (400 g). The young reach full size in 6–7 weeks, having been weaned when about a quarter grown. The life span of the pika is 1–3 years. During that time their predators are weasels and other small carnivores as well as hawks.

A modern guinea pig?

Pikas enjoyed a measure of obscurity for a long time, but they now seem to be emerging to fame as laboratory animals. They were first discovered in North America in about 1828. Naturalist Thomas Nuttal recorded how he heard, in the Rockies, "a slender but very distinct bleat, like that of a young kid or goat. But in vain I tried to discover any large animal around me." Finally he located the little animal "nothing much larger than a mouse." The first pika was discovered in Asia in 1769. However, little more was known of them, except their habit of storing hay and their ventriloquial voice, until the middle of the 20th century. Since then scientists in the former Soviet Union have been using them as laboratory animals because they are so easy to maintain.

PIKAS

CLASS **Mammalia**

ORDER **Lagomorpha**

FAMILY **Ochotonidae**

GENUS ***Ochotona***

SPECIES **26, including Pallas' pika, *O. pallasi*; and Rocky Mountain pika, *O. princeps***

ALTERNATIVE NAMES
Rock rabbit; mouse-hare; rock coney; calling hare; piping hare; whistling hare

WEIGHT
4⅖–14 oz. (125–400 g)

LENGTH
Head and body: 5–12 in. (12.5–30 cm)

DISTINCTIVE FEATURES
Resemble a small rabbit or hare, but with shorter, rounder ears and shorter legs; no visible tail; feet have hairy soles; grayish brown fur, lighter on underparts

DIET
Grasses, herbs, shoots and bark

BREEDING
Age at first breeding: 6 weeks; breeding season: varies with species and location; gestation period: 30 days; number of young: usually 2 to 5; breeding interval: up to 3 litters per year

LIFE SPAN
Usually 3 years, occasionally up to 7 years

HABITAT
Varied, includes open plains, deserts, mountains and steppes; usually on rocky terrain

DISTRIBUTION
Western North America and Asia

STATUS
Some species common (including both U.S. species); at least 4 Asian species threatened

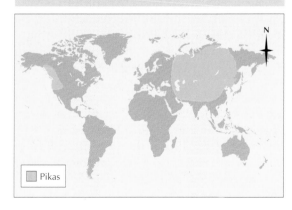

Pikas

PIKE

THE PIKE, APTLY NICKNAMED the "freshwater shark," is the fiercest predatory fish in the fresh waters of the Northern Hemisphere. It and its relatives, the redfin pickerel (*Esox americanus americanus*), grass pickerel (*E. americanus vermiculatus*) and muskellunge (*E. masquinongy)*, all of North America, are held in awe by some fishers, in contempt by others. To many they present a challenge, backed by the legends of size and ferocity—often very tall stories. The record of the largest pike caught is of a 53-pound (24-kg) specimen.

The pike, or northern pike as it is known in North America, is long-bodied with a large, flat, almost shovel-shaped head, large jaws and large mouth bristling with teeth. Its dorsal and anal fins are set far back. Its color ranges from olive to dark green with pale yellow spots. It grows up to 5 feet (1.5 m) long and weighs up to 77 pounds (35 kg). The muskellunge of the Great Lakes is very similar to the pike but has scales on only the upper part of the cheek instead of all over it. It grows to 8 feet (2.4 m) long and can weigh up to 110 pounds (50 kg). The chain pickerel, *E. niger*, found from Nova Scotia to Texas, grows up to 3 feet (90 cm) long and has dark bands on the flanks. The smaller redfin pickerel and grass pickerel of the eastern United States grow up to 14 inches (36 cm) long and have a chainlike network of dark markings on the sides. The black-spotted pike, *E. reichertii*, which is sometimes called the black-spotted pickerel, lives in eastern Siberia. Very little is known about this fifth member of the family.

Waiting in ambush

Pike live in still and running water, spending most of their time motionless among water plants, with which their colors harmonize. Camouflage is very important. On its back, head and upper sides the northern pike has irregular rows of light olive spots on a green-gold to light brown background. The belly is usually cream or white. The pike is therefore very well concealed among the vegetation at the bottom of a lake or river, where it lies in wait for passing prey.

The pike usually stay in one place and dart out to ambush their prey. Having the dorsal and anal fins set far back on the body gives great thrust to the tail and rapid acceleration, sending

With its long, torpedo-shaped form and its dorsal and anal fins set far back on its body, the northern pike can move very fast through the water after prey.

PIKE

CLASS	**Osteichthyes**
ORDER	**Esociformes**
FAMILY	**Esocidae**

GENUS AND SPECIES **Northern pike, *Esox lucius*; redfin pickerel, *E. americanus*; muskellunge, *E. masquinongy*; black-spotted pike, *E. reichertii*; chain pickerel, *E. niger***

ALTERNATIVE NAME
Redfin pickerel: grass pickerel (subspecies *E. americanus vermiculatus* only)

WEIGHT
Northern pike: up to 55–77 lb. (25–35 kg)

LENGTH
Northern pike: up to 5 ft. (1.5 m)

DISTINCTIVE FEATURES
Torpedo-shaped body; broad head; pointed snout; huge jaws and wide gape; small, backward-pointing teeth on upper jaw; large teeth on lower jaw

DIET
Adult: mainly other fish; also frogs, waterbirds, small mammals and crayfish

BREEDING
Breeding season: February–May; number of eggs: up to 500,000, laid in batches of 5 to 60; hatching period: 2–3 weeks

LIFE SPAN
Up to 30 years

HABITAT
Shady areas of lakes, pools, creeks and rivers

DISTRIBUTION
Northern pike: temperate regions of Northern Hemisphere; muskellunge: Great Lakes; grass and redfin pickerel: eastern U.S.; chain pickerel: Nova Scotia south to Texas; black-spotted pike: eastern Siberia

STATUS
Generally common

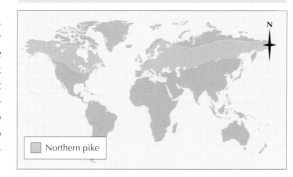

Northern pike

The northern pike has a huge, gaping mouth packed with teeth, which enable it to catch and hold onto large prey. It feeds mainly on other fish but will also seize frogs, ducklings, water voles and muskrats.

the pike out 20–30 feet (6–9 m) to seize prey. A pike detects its prey by sight rather than by smell, at distances of up to 50 feet (15 m) by day. However, a blind pike can also catch food and is probably warned of the approach of prey by the vibrations it makes in the water.

A pike can see at night as well as by day. Its habit is to lie well down in the water because its eyes are set in the top of the head and look mainly forward and upward. It has two sighting grooves, like gun sights, running to the tip of the snout. A pike's brain is relatively very small, 0.00075 of the total body weight, much of this being taken up by the optic lobes. This reflects the little effort the pike needs to make a living.

Steel trap jaws

When very small, pike feed on water fleas, worms and the fry (young) of other fish. As they grow, they take progressively larger fish and are less and less tempted by small prey unless it comes so close they can snap it up without moving. Adult pike are almost exclusively fish-eaters, especially of fish belonging to the carp family Cyprinidae and trout. Large pike will also take other water dwellers such as ducks, moorhens, gallinules, coots, water voles and frogs.

There are many authenticated reports of pike eating prey their own size. This is possible no matter how much the prey struggles because of the pike's teeth. Those teeth on the sides of the lower jaw are strong and stick straight up. They are used for seizing prey. The teeth of the upper jaw are smaller, most numerous in the front and curved backward. The roof of the mouth is bristling with backward-pointing teeth, which prevent prey from slipping out of the mouth. The mouth itself has a wide gape.

Sometimes large prey may become jammed in the pike's throat, with fatal results for the pike, which cannot get rid of it because of the backwardly directed teeth.

Large prey successfully swallowed takes a long time to digest. After a big meal the pike lies inert for a week or more, often near the bottom, taking no notice of prey or of a fisher's bait. A pike can take in large prey because its intestine is more or less straight, and its stomach is merely a dilation of the front part of the intestine.

Unusual digestive juices?

A hungry pike will seize prey of a certain size, depending on its own size and provided its quarry does not move too slowly, although an angler can catch pike using stationary dead bait. Pike learn, however, not to go after sticklebacks once they have experienced their sharp spines.

It is sometimes said that the pike's digestive juices are remarkably strong because even fish hooks are eaten away by its stomach acids. This is, in fact, illusory. When digesting a fish, the pike's acidity is high on the surface of the prey and very low inside it. A pike takes 3–5 days to digest a moderate-sized fish. A similar high acidity on the surface of a hook would soon erode it.

Many eggs wasted

Pike generally spawn from February to May, the younger individuals spawning first. They are stimulated to do so by the increasing day length and light intensity. Spawning generally occurs during the day, and the eggs are released in batches. The pike assemble in shallow water, each female usually attended by several males. Estimates of the number of eggs laid by each female vary from 40,000 to 500,000, the number probably depending on the size of the female. The males release sperm simultaneously. The fish may spawn for several hours, after which they rest before continuing, perhaps for several days until all the eggs are released.

During the resting period, both male and female pike may take new mates. They do not build nests or tend their eggs, which are simply scattered among the vegetation.

Many eggs are not fertilized because the micropyle, the hole in the egg membrane through which the sperm enter, closes 30–60 seconds after the eggs are laid. At first the eggs are sticky and lie singly on the bottom, later rising just off the bottom. They hatch after 2–3 weeks, the larvae feeding on the remains of the yolk sac for 10 days before starting to catch their own food. The parents take no care of their young, which must find protection among the dense vegetation.

Automatic control of numbers

As pike are at the apex of a food pyramid, they have few predators except when very young. There is, however, considerable cannibalism, which keeps a proper balance. The more richly a water is stocked with other fish, the less often this cannibalism occurs. The end result is that pike are seldom so numerous as to deplete the waters in which they live of other fish.

Most pike live about 7 years if they survive the massacre of infancy but they have been known to live much longer. C. Tate Regan once asserted that "fish of sixty or seventy pounds weight are at least as many years old."

Pike are solitary fish and tend to stay close to home. They usually lurk in weed beds and in the shady waters of larger lakes, although they also live in pools and in slow-moving rivers and streams.

PILCHARD

THE NAME PILCHARD REFERS to five species of fish belonging to the family Clupeidae, including the European pilchard, *Sardina pilchardus*, and the South American pilchard, *Sardinops sagax*. Pilchards are highly exploited by people, as is their relative the herring, *Clupea harengus*. An adult pilchard is a silvery fish a little smaller and slightly fatter than a herring. It grows up to 15½ inches (40 cm) long in the case of the South American pilchard and up to 10 inches (25 cm) long in the case of the European species. Pilchards resemble the herring in appearance except that a pilchard's dorsal fin is slightly forward of the midline of the body, whereas that of the herring is more or less at the center. Also, a pilchard's scales are larger than a herring's but share with them the quality of being deciduous. That is, they lie in shallow pockets on the surface and are easily rubbed off.

Young are called sardines

Young pilchards are known commercially as sardines. For example, young European pilchards support an extensive "sardine" canning industry in France, Spain and Portugal, while an adult "pilchard" fishery formerly flourished off Cornwall, England. This division into sardine and pilchard arose because the European pilchard has several geographic races and makes extensive migrations. The European pilchard can be divided into northern, southern, Moroccan and Mauretanian subspecies, each with its own distinct geographical distribution.

The South American pilchard lives in the Agulhas, Benguela, California, Kuroshio and Peru currents, and off Australia and New Zealand. The Australian pilchard, *Sardinops neopilchardus*, is found off the southern half of Australia as well as around New Zealand, and the genus also contains the Japanese pilchard (*S. melanostictus*) and the southern African pilchard (*S. ocellatus*).

Schooling by age

Pilchards live in large schools made up of fish of approximately the same age and size. The depths at which they swim vary according to the time of day. The fish tend to follow the vertical migrations of plankton, which move deeper by day and rise toward the surface at night.

Pilchards pick off large numbers of copepods and other planktonic crustaceans by means of very fine gill rakers. They use the gill rakers to sieve even small diatoms from the water as it passes into the mouth and throat and out again by the gills. European pilchards feed most from April to July and in October, when they are at their heaviest. From November to February or March pilchards fast, eating little, if anything. At this time they can become quite thin.

PILCHARDS

CLASS	**Osteichthyes**
ORDER	**Clupeiformes**
FAMILY	**Clupeidae**
GENUS	**Sardina** and **Sardinops**
SPECIES	**5, including European pilchard, Sardina pilchardus; and South American pilchard, Sardinops sagax**

ALTERNATIVE NAME
Sardine (young only)

WEIGHT
Up to 1.1 lb. (0.5 kg)

LENGTH
European pilchard: up to 10 in. (25 cm), usually less; South American pilchard: up to 15½ in. (40 cm)

DISTINCTIVE FEATURES
Herringlike form; dorsal fin originates in front of pelvic fins; blue green above, shading to gold on sides; whitish elsewhere; up to 4 dusky spots along sides at eye level

DIET
Mainly planktonic crustaceans

BREEDING
Breeding season: spring and summer

LIFE SPAN
Up to 25 years

HABITAT
Coastal waters

DISTRIBUTION
European pilchard: northeast Atlantic south to northwestern Africa; also Mediterranean and Black Sea. South American pilchard: Pacific seaboard of Americas; Japan south to Australia and New Zealand.

STATUS
Abundant

South American pilchard   European pilchard

The European pilchard spawns from April to July, when each female lays up to 60,000 large eggs. Females spawn 20–40 miles (32–64 km) from shore in waters with a temperature of 50–63° F (10–17° C). Pilchard eggs are unusual in having a large space between the egg itself and the outer membrane. This characteristic, together with the large globule of oil each contains, makes pilchard eggs buoyant, and they float at the surface. The larvae hatch in 3 days or so, each being 4 millimeters long.

Although herringlike in general form, pilchards have larger, easily detached scales and a rounded body. These are young pilchards, or sardines, displayed on ice in Venice, Italy.

Pilchard migrations

The exact temperature needed for spawning differs for each race of European pilchard but lies between the limits quoted above. The northern race has two distinct populations. One spawns in the English Channel. The young fish then migrate south along France's Atlantic coast, returning 2 years later to spawn on the grounds where they hatched. They are fished as sardines off the western coast of France and the northern coast of Spain, and as pilchards off southwest England. The second population spawns in the Bay of Biscay, the young fish spending their first 2 years in the same place, after which they migrate north as mature fish and do not return to their birthplace. In the remaining two races of the European pilchard, young fish and mature fish also occupy separate ranges

Migrations of mature European pilchards take place as follows: from Gibraltar north to Galicia in northern Spain (November–June); from Santander in Spain north to Arcachon in France (November–April); off Brittany in northwestern France (February–July) and off southwest England (April–November). The regular appearance from south to north of schools of adult pilchards gave rise to the idea that there was an extensive south–north migration, whereas it is a succession of migrations of local populations.

PILOT FISH

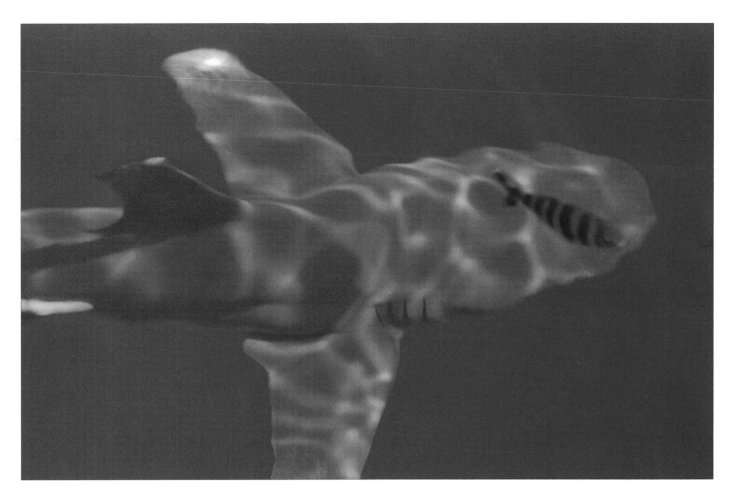

A pilot fish swims alongside a whitetip shark, Carcharhinus longimanus. *It is still a mystery why pilot fish accompany large hosts. The most likely reason is simply that they are are hitching a ride.*

THE PILOT FISH IS so named because it was believed to guide sharks and whales and to lead ships to a port or solitary swimmers to land when they had lost their way. These beliefs go back at least to the time of the ancient Greeks. The pilot fish can grow up to 27½ inches (70 cm) long. It has a strongly forked tail that is blackish with white tips, a prominent dorsal fin with four strong spines in front of it, a prominent anal fin and small pectoral and pelvic fins. The pilot fish's body is marked with five to seven dark gray-blue bands on a background of variable color. The species is widespread through tropical and temperate seas.

Pilot fish spend their lives swimming in company with other ocean creatures, such as whales, large sharks, giant devilfish (genera *Manta* and *Mobula*) and schools of tuna, and even with sailing ships. They have sometimes been caught in herring and mackerel nets. Although the ancient belief is that pilot fish lead, and therefore guide, they more commonly swim at the side of their hosts, or even follow them. Nevertheless, the association between the pilot fish and its host is a very persistent one. Pilot fish accom-

panying a shark that has been hooked and hauled on board ship have swum around their ensnared companion's tail, the last part of it to leave the water, as if distracted.

Female pilot fish spawn well away from land, and the eggs float at or near the surface. Young pilot fish have large eyes and numerous spines on the head. They shelter under the bells of jellyfish, amid the tentacles of Portuguese men of war (genus *Physalia*) or under bunches of Sargasso weed (genus *Sargassum*) and pieces of floating wreckage. It is supposed that by swimming in company with jellyfish, the young pilot fish are protected from predators. This may be true, but they are also exposed to the dangers of being eaten by animals that feed on these organisms, including seabirds such as frigate birds (family Fregatidae) and fulmars (genus *Fulmarus*) and fish, including the ocean sunfish, *Mola mola*.

Why follow sharks?

Why pilot fish accompany large animals and other objects remains something of a mystery. It was believed that pilot fish guided sharks and rays toward suitable prey, receiving in return

PILOT FISH

CLASS **Osteichthyes**

ORDER **Perciformes**

FAMILY **Carangidae**

GENUS AND SPECIES **_Naucrates ductor_**

LENGTH
Up to 27½ in. (70 cm)

DISTINCTIVE FEATURES
Torpedo-shaped body, rather rounded in cross section; strongly forked, blackish tail with white patch near each corner; dark gray blue coloration on back and top of head, continuing as 5 to 7 broad bands down sides and onto belly

DIET
From water near to host: small fish and invertebrates. From host itself: scraps of leftovers, parasites and excrement.

BREEDING
Poorly known

LIFE SPAN
Not known

HABITAT
Seas and oceans. Adult: extremely close relationship with sharks, rays, large bony fish and turtles. Young: under bells of jellyfish, bunches of Sargasso weed and pieces of floating wreckage.

DISTRIBUTION
Tropical and subtropical waters worldwide

STATUS
Not threatened

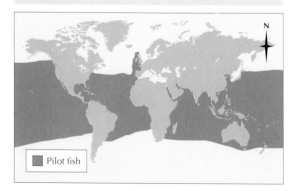

Pilot fish

A group of pilot fish swimming near the entrance to the Arabian Gulf, the dark vertical bands on their bodies clearly visible.

feeding forays of their large companions, on the few occasions that pilot fish have been caught and their stomach contents examined, they seem to have been eating small fish rather than just scraps of larger prey.

There is also uncertainty about whether protection is a factor in the relationship between pilot fish and host. It is usually taken for granted that the pilot fish is afforded protection by a shark host because sharks are so voracious that potential predators are unlikely to come near. However, a pilot fish seems unlikely to get protection in return for accompanying a whale or a shoal of tuna, and less likely still for sheltering beneath a bunch of Sargasso weed. It seems improbable that the benefit of protection holds for one type of host and not for the others.

Life in the slipstream?

The studies of the Soviet scientist V. V. Shuleikin, published in the 1950s, give one valid reason for pilot fish swimming beside sharks. Shuleikin calculated that sharks swim three times as fast as a pilot fish can. How then does a pilot fish keep up? Over the surface of any body moving through water there is a boundary layer of water moving forward at almost the same speed as that of the body. This layer is thickest over the tail half of a shark, which is where a pilot fish usually swims. Presumably the pilot fish is able to travel hundreds of miles with a minimum of effort, carried along in this boundary layer.

protection from predators because of their proximity to a formidable companion. This notion dates at least from the 16th century, when pilot fish were called sharks' jackals. In reality both pilot fish and their shark hosts are in search of food. Although pilots no doubt benefit from the

PILOT WHALE

Short-finned pilot whales, such as this one off Hawaii, prefer warmer waters than the long-finned species. The short-finned variety also have shorter flippers and fewer teeth.

ALSO CALLED THE BLACKFISH or pot-head whale, the pilot whale is a dolphin that can grow more than 21 feet (6.5 m) long. Its most distinctive feature, from which it gets its scientific name of *Globicephala,* is its bulbous forehead. This forms a dome overhanging the upward curved mouth. The pilot whale is also known for its strong "blow" when exhaling. The tapered flippers are long and the hooked dorsal fin stands about 1 foot (30 cm) high. Pilot whales are black or dark gray in color.

Pilot whales are found offshore and in oceans in many parts of the world but are absent from the polar seas. There are two species: the long-finned pilot whale, *Globicephala melas,* and the short-finned pilot whale, *G. macrorhynchus.* The first has longer flippers, as its name suggests, and tends to be found in cooler waters than the short-finned species. It ranges as far north as Greenland and is still fairly common around the Faeroe Islands, off Norway and around the islands of Shetland and Orkney, north of Scotland.

Follow the leader

Pilot whales live in large schools, but after heavy exploitation by hunters, these rarely reach the hundreds of individuals that used to be a common sight. Each school is made up of males and females and their young. It has often been reported that a female acts as a leader. This is presumably how these animals got their name, although it is more likely that there is a general tendency to follow any individual.

The habit of following a leader has, however, made schools of pilot whales susceptible to mass strandings. If one whale becomes confused and goes ashore, the others in the school will often follow it. Where this happens, attempts are often made to refloat the stranded pilot whales, but these usually end in failure. As soon as one whale is towed into deeper water, it swims back to the shore. Presumably the rescued whales are blindly answering the calls of the stranded ones. Orcas, *Orcinus orca,* also known as killer whales, sometimes behave in the same way.

PILOT WHALES

CLASS   **Mammalia**

ORDER   **Cetacea**

FAMILY   **Delphinidae**

GENUS AND SPECIES   **Long-finned pilot whale,** *Globicephala melas;* **short-finned pilot whale,** *G. macrorhynchus*

ALTERNATIVE NAMES
Blackfish; pot-head whale

WEIGHT
⁹⁄₁₀–3⅗ tons (1–4 tonnes)

LENGTH
11⅘–21⅓ ft. (3.6–6.5 m)

DISTINCTIVE FEATURES
Bulbous forehead; upward curved mouth; hooked dorsal fin; black or dark gray. Short-finned pilot whale: shorter flippers and fewer teeth than long-finned species.

DIET
Mainly squid and cuttlefish; also fish

BREEDING
Age at first breeding: 7–12 years (female), 15–22 years (male); breeding season: peaks October–November (Northern Hemisphere); number of young: 1; gestation period: 15 months; breeding interval: about 7 years

LIFE SPAN
Up to 45 years

HABITAT
Offshore waters and open oceans; short-finned pilot whale prefers warmer waters to long-finned species

DISTRIBUTION
Most parts of the world, excluding polar seas

STATUS
At low risk; some subspecies conservation dependent

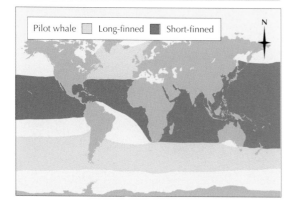

Pilot whale   ☐ Long-finned   ■ Short-finned

Why an individual whale will swim into shallow water in the first place, taking the rest of the school with it as it does, remains a mystery. Certainly if a herd swims down an estuary, it will find it difficult to find its way back to the open sea. Some biologists believe that such frequent strandings probably have a significant effect on pilot whale numbers, although some populations remain reasonably abundant at present.

A female pilot whale and her calf. Each female gives birth to only a single calf every 7 years, on average.

Sonar hunting

Pilot whales are armed with 14 to 24 teeth in each side of the mouth, used for grabbing slippery prey. They feed mainly on cephalopods such as squid and cuttlefish, but fish are also taken. The eyesight of pilot whales is not good, and their bulbous heads and inflexible necks probably prevent them from seeing objects directly ahead. However, like other dolphins, they mainly use sonar (a method of emitting and receiving sound waves so as to detect and locate objects underwater) to hunt for food.

Warm-water breeding

Studies of schools of pilot whales have shown that they migrate regularly. The migrations are partly regulated by food supply, as when pilot whales move to the coast of Newfoundland in summer after the squid *Illex,* but there is also a more general movement to warmer waters in winter.

In general, mating takes place in warmer waters in winter, but varies between different locations and schools. In the Northern Hemisphere breeding peaks in October and November.

The single calf is born after a gestation period of 15 months, when the pilot whales return to cooler regions the following year.

It is usual for whales such as the blue whale, *Balaenoptera musculus*, to migrate to warmer seas to breed because newborn whales have very little insulating blubber. To compensate for this, the female whale's milk is extremely rich, comprising 40–50 percent fat. This compares to cow's milk, which is just 4 percent fat.

Until recently very little was known about the mating habits of whales. However, pilot whales have been seen courting in the wild. Often courtship will take place between several pairs in a school simultaneously. The male and female stroke each other with their flippers or bodies as they swim slowly past. They also swim side by side, playfully biting one another's mouths, then surface vertically until their flippers are exposed. Eventually the two submerge and lie belly to belly for about 20 seconds; copulation in whales being extremely rapid.

Female pilot whales start to breed when between 7 and 12 years old and bear young on average only every 7 years. Males mature much later, at 15–22 years. Pilot whales have been known to live for a maximum of 45 years.

Wholesale slaughter

For hundreds of years pilot whales were hunted around the islands to the north of the British Isles. These hunts used to depend on a school being sighted near the shore. The alarm was then given and a fleet of small boats would put out to shepherd the school into a bay. As the whales panicked the school would become stranded in the shallows and on the beach. Then people from the boats and groups of people waiting on the shore would set on the whales with knives and lances, slaughtering the entire group: males, females and their young.

These hunts have now ceased in the Orkneys and Shetlands, but continue to take place in the Faeroe Islands. The slaughter of a school of pilot whales used to be a windfall to the islanders, and even now there are wild scenes following the killing of *grindehval,* as the Faeroese call them. However, the hunting of these animals has become increasingly controversial, in the Faeroes and elsewhere where the animals are still exploited. While some populations seem to stay stable, others have crashed. Some subspecies are now classed as conservation dependent, while both the long-finned and short-finned species are said to be at low risk.

A school of long-finned pilot whales, Shetland. After decades of exploitation by hunters, often involving the slaughter of entire schools, these whales are not nearly so common as they once were.

PINK-FOOTED GOOSE

THE PINK-FOOTED GOOSE IS one of the so-called gray geese that together make up the genus *Anser*. Other gray geese include the greylag goose, *A. anser*, discussed elsewhere, and the bean goose. The latter is so named because it arrives in England in October, at the time of the bean harvest, and stays to feed on the beans left lying in the fields. Bean geese have been the subject of intense study by ornithologists in the last few years of the 20th century. The current view, held by most experts, is that there are two species of bean geese. The taiga bean goose, *A. fabalis*, has a long, heavy bill and is larger than the tundra bean goose, *A. serrirostris*. As their names suggest, these species breed in different climatic zones.

The pink-footed goose is 24–29½ inches (60–75 cm) in total length. The head and neck are dark brown, the underparts are light pinkish brown and the wings and tail are ash gray with white edgings. The feet are pink and so is the bill, except the base and tip, although it may very occasionally be wholly pink.

Bean geese breed in the tundra and taiga forest zones, from Greenland to eastern Siberia, but the pink-footed goose is restricted to eastern Greenland, Iceland and Spitzbergen. In the winter pink-footed geese migrate to Britain, northern France, Belgium, the Netherlands and Germany. They occasionally turn up in North America, Russia and other parts of Europe.

Wary geese

During the winter pink-footed geese gather in very large flocks on sandbanks, moors, estuaries, flooded marshes and coasts, places where they are unlikely to be disturbed. Pink-foots are extremely wary and difficult to approach. They arrive in their winter quarters in September and October, having migrated from the breeding grounds, with only a few stops, in the Faeroe Islands or Scandinavia. They fly in skeins of over 1,000 geese, with family parties of adults and goslings keeping together. When they arrive, weary and hungry, they are less wary than usual and for a few days it is possible to get nearer to them before they recover their strength. The flocks stay south until April or May.

Grazing and gleaning

In the fall and winter, pink-footed geese feed in stubble fields, usually of barley, but they also eat young wheat and grass stems. On the breeding grounds they eat the buds, stalks, seeds and leaves of many plants, including willows, sedges,

horsetails, chickweed and grasses. These days, winter flocks feed mainly in arable fields, on barley, oats, wheat, potatoes and carrot roots.

Safe nesting places

By the time the pink-footed geese have reached the breeding grounds they have already paired off. After only a few days the eggs are laid. Most pairs nest in loose colonies, but some raise their broods singly, particularly in the open tundra. If large numbers nested together in open country, predators such as Arctic foxes would be attracted to such easy prey. The colonies are usually in more inaccessible places such as the talus, the piles of boulders that form at the bottom of cliffs, on moraines, cliff ledges or terraces. Some colonies are on islands in rivers. When on level ground, the nests are built on hillocks or frost mounds where the sitting bird can get a good

A wary pink-footed goose incubates her eggs. She spends around 4 weeks sitting on the nest while the gander (male) acts as a lookout for predators such as Arctic foxes and black-backed gulls.

view of the surrounding country. The gander (male) usually keeps a watch from a lookout point nearby.

The nest is a depression in the ground lined with grasses and other plants and a considerable quantity of down. The three to five eggs, sometimes more, are incubated by the goose (female) alone while the gander stands guard. The eggs hatch in just under 4 weeks. The chicks leave the nest within 48 hours and never return. Their parents lead them down to water when they leave the nest. A new roost is used each night.

The adults molt during the breeding season, shedding all their feathers simultaneously, so they cannot fly until new ones grow. As soon as they and their young can fly, they migrate south.

When they have molted their flight feathers, the geese are helpless. Although their natural wariness usually keeps them safe, they will fall prey to Arctic foxes unless they are in an isolated place. Greater black-backed gulls (*Larus marinus*), gyrfalcons (*Falco rusticolus*) and white-tailed eagles (*Haliaeetus albicilla*) also prey on the eggs and chicks of pink-footed geese in Iceland.

Goose round up

The Arctic breeding grounds of pink-footed geese were discovered in Spitzbergen in 1855 and in Greenland in 1891. These colonies were fairly small, however, and the main colonies in Iceland were discovered only in 1951, when the ornithologists Peter Scott, James Fisher and Finnur Gudmundson found over 2,000 nests. They banded over 1,000 pink-footed geese to find where they spent the winter, catching them by driving them into nets. At that time of the year neither adults nor goslings could fly, so they were easy to capture. Nearby they found the remains of U-shaped stone walls, which were the ruins of goose pens that had been used by Icelanders centuries before for rounding up flightless geese for food.

Flocks of pink-footed geese feed in arable fields in winter. They favor barley, oats, wheat, potatoes and carrots.

PINK-FOOTED GOOSE		
CLASS	**Aves**	
ORDER	**Anseriformes**	
FAMILY	**Anatidae**	
GENUS AND SPECIES	***Anser brachyrhynchus***	

WEIGHT
4–6⅗ lb. (1.8–3 kg)

LENGTH
Head to tail: 23½–29½ in. (60–75 cm); wingspan: 54–67 in. (1.35–1.7 m)

DISTINCTIVE FEATURES
Compact goose; dark brown head and neck; pale gray underparts with pinkish tone; ash-gray wings and tail; pink bill and legs

DIET
Summer: buds, shoots, stalks, seeds, leaves, roots and berries of many plants; winter: cereals, carrots, potatoes, roots and grasses

BREEDING
Age at first breeding: usually 3 years; breeding season: eggs laid late May–June; number of eggs: usually 3 to 5; incubation period: 26–27 days; fledging period: about 56 days; breeding interval: 1 year

LIFE SPAN
Up to 20 years

HABITAT
Summer: river valleys, grassy slopes and heathy uplands in tundra zone; winter: lowland arable fields; roosts on freshwater lakes and estuary mudflats

DISTRIBUTION
Summer: 2 breeding populations, in Greenland/Iceland and Spitzbergen; winter: British Isles, Ireland, Netherlands, Denmark and westernmost Germany

STATUS
Common but highly localized

Pink-footed goose ☐ summer ☐ winter

PINTAIL

THE NORTHERN PINTAIL, *Anas acuta*, is one of the most numerous ducks in the world. It is the same length as the mallard, *A. platyrhynchos*, 20½–26 inches (51–66 cm), except that the central pair of tail feathers are elongated. These form the pintail that has given rise to the bird's common name and which may grow to 4 inches (10 cm) long in breeding males.

The plumage of the male is distinctive, with a chocolate brown head and neck and a white collar running down the side of the neck and across the breast. The back is patterned and there is a creamy yellow patch just in front of the tail. The female is very much like a female mallard in appearance. The male goes into eclipse, that is, its plumage grows duller, from mid-July to early September. The eclipse plumage is like that of the female but is darker on the upperparts. The speculum (a patch of color on the secondary feathers of most ducks and some other birds) is bronzy green and a buff bar on the front is visible on the wings. The northern pintail is distributed around the Northern Hemisphere and migrates south in winter. Like the mallard, it is truly cosmopolitan; it has a huge range and is an extremely successful species.

The white-cheeked or Bahama pintail, *A. bahamensis*, ranges from the Bahamas south through most of South America and also breeds on the Galapagos Islands. It has distinctive white patches on the sides of its face. Some pintails breed farther south than any other type of duck. There are subspecies of the northern pintail on the subantarctic islands of Kerguelen and the Crozets, and South Georgia has its own separate species, *A. georgica*. Somewhat confusingly, these subantarctic pintails are known as teal.

Tundra nesters

The vast inhospitable wastes of the Arctic tundra provide the main breeding ground for the northern pintail, which nests in the numerous scattered pools and lakes there. Freshwater pools are preferred to brackish ones, and nests are not usually found on water surrounded by marshy ground. By contrast, in the Galapagos there are very few stretches of fresh water where the white-cheeked pintail can live. Most white-cheeked pintails occur in South America, in a variety of freshwater habitats. In Europe, northern pintails prefer wintering around coasts and estuaries, whereas in India large numbers

The northern pintail (male, above) is both widespread and abundant. It favors open wetlands such as marshes, lakes, estuaries and coasts.

Although they often feed at the water's surface, pintails also up-end themselves to feed underwater. Pictured is a Bahama or white-cheeked pintail.

are found on inland lakes. This is unusual because northern pintails usually only gather in small numbers on inland waters. On the sea they may form flocks of over 1,000 birds. These large flocks are split up into small parties that mingle with other ducks rather than forming tightly packed rafts.

Pintails feed by paddling in the shallows, up-ending, or by uprooting water plants such as pond weed, sedges and docks. Acorns and grain from stubble, beetles, fly larvae, worms and tadpoles are also taken. Their winter diet includes seaweed and eelgrass. Pintails only dive to escape danger when they are unable to fly. Such a situation may arise when they are molting during eclipse or because of wounds.

Courtship chases

The display courtship of the northern pintail is similar to that of the mallard and there are aerial pursuits in which the female is chased by several males. Pintails often perform aerial dives, plunging at an angle of 45 degrees from a great height, stooping with their wings stiffly outspread and slightly curved down. At the last moment they level out and glide a few yards off the ground for 100–300 yards (90–275 m).

The greatest concentrations of breeding northern pintails are probably in the tundra of Alaska, where every small pool houses one or more pairs. Elsewhere the nests are fairly close together, sometimes a long way from water. These are no more than a down-lined hollow in the ground among marram grass, heather, rushes or other low plants. The seven to nine eggs are incubated by the female and usually guarded by the male. The chicks hatch out after 22–24 days

NORTHERN PINTAIL

CLASS	**Aves**
ORDER	**Anseriformes**
FAMILY	**Anatidae**
GENUS AND SPECIES	***Anas acuta***

WEIGHT
1⅓–2½ lb. (0.6–1.1 kg)

LENGTH
Head to tail: 20½–26 in. (51–66 cm); wingspan: 32–38 in. (80–95 cm)

DISTINCTIVE FEATURES
Slender duck with thin neck. Breeding male: chocolate brown head and neck; pure white neck stripe, joining white underparts; gray upperparts; creamy yellow patch in front of tail; very long black central tail feathers. Female and nonbreeding male: drab grayish brown all over.

DIET
Leaves, stems and seeds of water plants; grain; small invertebrates and tadpoles

BREEDING
Age at first breeding: 1–2 years; breeding season: April–August; number of eggs: usually 7 to 9; incubation period: 22–24 days; breeding interval: 1 year

LIFE SPAN
Up to 25 years

HABITAT
Summer: shallow pools and lakes in Arctic tundra and lowland grassland. Winter: estuaries and sheltered coasts; also inland lakes and farmland.

DISTRIBUTION
Breeds throughout north of Northern Hemisphere; winters in temperate and tropical zones to south

STATUS
Very common

Northern pintail ☐ summer ☐ winter

and shortly afterward the parents lead them to the nearest source of water to feed. If they are disturbed, the chicks hide among water plants. There are records of a mother pintail performing distraction displays in which she comes close to the intruder and splashes about vigorously, swimming in circles. The young fly when they are about 6–7 weeks old.

Aerial highways

Ducks of the genus *Anas*, to which the pintail and mallard belong, make up three-quarters of the quarry of wildfowlers. In the United States in particular, vast numbers of wildfowl are shot every year as they fly across the country to and from the breeding grounds in the north. There are stringent game laws in place to protect the waterfowl and conservation agencies regularly prepare suitable stretches of water for the birds to settle. The agencies' task is made easier because waterfowl, in common with other birds, travel along traditional migratory routes.

Migrating birds usually follow the least demanding routes when they are traveling. Birds of prey, for instance, cross the Mediterranean by the shortest sea routes, over the Straits of Gibraltar and the Bosporus in Turkey. Sometimes, however, the only way is over difficult ground.

Northern pintails migrating in and out of India have to cross mountains, and the skeleton of one was found 16,000 feet (4,800 m) up Mount Everest. In North America there are four well-defined migration routes that follow geographic features. These were discovered and named flyways in the early 1950s by waterfowl biologist Frederick Lincoln, based on his studies of banded birds, mainly waterfowl.

The flyways start in the tundra of Alaska and northern Canada and continue down through the plains of Canada and into the United States. On the way, the ducks and geese from the Arctic are joined by hosts of other birds. They finish on winter grounds around the Caribbean Sea or in South America. The most important flyway follows the Mississippi valley south to the marshy shores of the Gulf of Mexico. Next in importance is the central flyway, also starting in Alaska but subsequently running close to the eastern side of the Rocky Mountains. Along the coasts there are the Atlantic and Pacific flyways, one leading to the Caribbean and South America, the other to the Pacific coast of Mexico and California. These flyways are less used because the winter climate of the coast is less severe than in the center of the continent. Therefore, there is less pressure on the birds to travel south.

Their dull brown plumage distinguishes female northern pintails (right, below) from breeding males (left, below). However, the plumage of young pintails and postnesting males is very similar to that of the females.

PIPEFISH

floridae, is normally dark green when among eelgrass but turns light when among pale green weeds. Other species of pipefish in American seas are muddy brown but turn brick red when among red weeds.

Most pipefish are marine or estuarine, but a few species are found in fresh water. The marine species live chiefly inshore, in shallow seas, but some live at depths of 50 feet (15 m) or more, and one lives among the weeds of the Sargasso Sea off the Florida coast.

Vertical swimmers

Pipefish in shallow waters and estuaries often live among eelgrass, the only flowering plant in the sea. They swim in a vertical position, with the dorsal and pectoral fins vibrating in time with each other, driving the fish through the water in a leisurely fashion. The vibrations of the dorsal fin are so rapid as to give the impression of a tiny propeller. Pipefish can also slip through the water with snakelike movements of the body, and turn their heads from side to side, using these movements for steering. The eyes can be moved independently of each other, as in chameleons. Pipefish may also be found among seaweeds, at times in rock pools, in holes and crevices, and there are species in tropical seas that live in the interstices in coral rock rubble, almost like earthworms.

Prey gets sucked in

Pipefish have no teeth and their jaws are permanently locked, supporting a tubular mouth. They cannot pursue prey, but the mouth acts almost like a syringe, sucking in small planktonic animals from as much as 1½ inches (4 cm) away. When searching for food, pipefish may swim upright or in a horizontal position, wriggling and twisting the body, turning the head and thrusting among tufts of weeds or into cracks and crevices. Pipefish seem to be selective, apparently scrutinizing small crustaceans such as copepods, tasting each morsel and rejecting it if not satisfied.

Male carries the young

The main scientific interest of pipefish is their breeding. They are related to sea horses, the males of which have a pouch in which the female lays her eggs. In pipefish reproduction is more

A multibar pipefish, Dunckerocampus multiannulatus, in the Red Sea. Pipefish belong to the same family as sea horses and resemble them in some aspects, including reproduction.

THESE EEL-LIKE FISH WERE GIVEN their name in the mid-18th century, when pipe stems were long and very thin. The shape of the body has led to common names such as worm pipefish, snake pipefish and threadfish. There are about 215 species belonging to some 52 different genera. They range in size from 1 inch to 1½ feet (2.5–45 cm) and are found in both tropical and temperate seas. All are long and very slender, with long heads, tubular snouts and tufted gills. Instead of scales, pipefish have a series of jointed, bonelike rings encircling the body, from behind the head to the tip of the tail. Some species have a small tail fin. There is also a dorsal fin positioned at midbody.

The colors of pipefish are usually tan, greenish or olive, like the seaweed among which they live. Most species have a slight banded pattern, which is particularly well marked in banded pipefish such as the ringed pipefish, *Doryrhompus dactyliophorus*, and the network pipefish, *Corythoichthys flavofasciatus*. Some species are mottled and spotted, while others are able to change color. For example, the dusky pipefish, *Syngnathus*

BAY PIPEFISH

CLASS	**Osteichthyes**
ORDER	**Syngnathiformes**
FAMILY	**Syngnathidae**
GENUS AND SPECIES	***Syngnathus leptorhynchus***

LENGTH
Up to 1 ft. (30 cm)

DISTINCTIVE FEATURES
Long, slender body covered by 17 to 22 bony rings (scales absent); tubular snout of moderate length; tufted gills; dorsal fin at midbody; tan, sometimes greenish, in color

DIET
Small crustaceans such as copepods

BREEDING
Age at first breeding: not known; breeding season: May–June; number of eggs: up to 225; hatching period: 3–4 months; breeding interval: not known

LIFE SPAN
Not known

HABITAT
Mainly among eelgrass in bays and estuaries; sometimes in shallow, offshore waters

DISTRIBUTION
Eastern Pacific: Sitka, Alaska, south to southern Baja California, Mexico. Northern population: Alaska south to Monterey Bay. Southern population: Morro Bay southward.

STATUS
Common

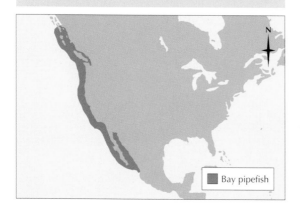

Bay pipefish

A greater pipefish, Syngnathus acus, England. Pipefish are unusual in that the body is covered by a series of jointed, bony rings rather than scales.

For example, in some species there are merely two folds of skin in which the eggs lie. All pipefish have, however, one thing in common: the male carries the offspring. Also, in many species the female does all or most of the courting.

Submarine flirtation

The female broadnosed pipefish, *S. typhle*, of the Mediterranean, for example, courts the male for several hours before he responds by swinging his body from side to side, through a right angle. The female will go on to court one male after another, and may even mate with several in turn.

Courtship in the dusky pipefish has been studied in greater detail. First the male and female swim around each other in the vertical position. They swim in decreasing circles until their bodies touch, at which point the male bends farther forward and caresses the female with his snout. Finally, their bodies become entwined, and the female inserts her ovipositor into the male's pouch and lays some eggs. The male wriggles his body to work the eggs down into the pouch, after which the female lays more eggs, and this is repeated until the pouch is full. The eggs hatch and the larvae are shot out, one or a few at a time, by the male making convulsive movements. Even when able to swim freely, the baby pipefish may dive back into the pouch in times of danger.

In the bay pipefish, *S. leptorhynchus*, mating takes place in May and June. The female deposits up to 225 eggs in the male's pouch. Studies have found two or three distinct developmental stages of eggs in a male's pouch, which could have been from different female fish. The males usually carry the eggs in May and then continue to carry the hatched young in August. However, eggs and larvae have been found in the male pouch from February through November in this species.

complicated. In some species the female merely lays her eggs on the underside of the male, where they stick. At the other extreme are species in which the male has a long pouch formed by folds of the surface growing down and meeting in the middle. There are other pouch types as well.

PIPISTRELLE

THE COMMON PIPISTRELLE IS both the most familiar and the smallest of European bats, although there are several other species of pipistrelles in continental Europe that are only slightly larger. What used to be thought was a single species in Britain has now been revealed to be two: the common pipistrelle, *Pipistrellus pipistrellus*, the most common bat in Britain, and the soprano pipistrelle, *P. pygmaeus*. The common pipistrelle, like most small bats, has a jerky, erratic flight and is recognizable because it appears earlier in the evening than most other bats, sometimes even coming out in broad daylight. The jerky flight of pipistrelles gave rise to the ancient name for bats, which was flitter-mouse, *Fledermaus* in German. Another medieval name for bats was reremouse.

Weighing only 1/10–1/5 ounce (3–6 g) and with a maximum head and body length of 2 inches (5 cm), the common pipistrelle is nevertheless of robust build, with short legs, a broad, flat head and a 1-inch (2.5-cm) tail. The interfemoral membrane stretches between the thighs and the tail, but does not extend to the last tail joint. The wingspan of the common pipistrelle is 8–9 inches (21–23 cm). The calcar, a spur on the hind leg,

reaches almost to the tail, and behind it is a lobe of membrane, known as the postcalcareal lobe, on the interfemoral lobe.

Except on the ears and wing membranes, where it is blackish, the somewhat silky fur of the bat's upperparts varies from dark to light reddish brown. The underparts are paler and the undersurface of the interfemoral membrane has no fur. The common pipistrelle's muzzle is blunt with a wide mouth, and the short, somewhat triangular ears are slightly notched on their outer edges and just behind the angle of the mouth.

Altogether there are about 77 species of pipistrelles. The range of the common pipistrelle includes the whole of Britain and much of continental Europe. It is possible that its range also extends to southwestern Asia, northeastward to Korea and Japan and eastward to Kashmir, as well as to Morocco, although there is currently still scientific debate on this point. Other species live in Madagascar, the Malay Archipelago, Australia and the Philippines. There are two species in North America. The eastern pipistrelle, *P. subflavus*, is found throughout the eastern United States, while the western pipistrelle, *P. hesperus*, occurs in the west of the country.

The common pipistrelle bat, seen here flying between birch catkins, is one of 77 species of the Pipistrellus *genus. Pipistrelle bats are found worldwide, including two species in North America.*

COMMON PIPISTRELLE

CLASS **Mammalia**

ORDER **Chiroptera**

FAMILY **Vespertilionidae**

GENUS AND SPECIES **_Pipistrellus pipistrellus_**

WEIGHT
⅒–⅕ oz. (3–6 g)

LENGTH
**Head and body: up to 2 in. (5 cm);
wingspan: 8–9 in. (21–23 cm);
tail: about 1 in. (2.5 cm)**

DISTINCTIVE FEATURES
**Very small size; robust build with short legs;
broad, flat head; small ears with earlets;
interfemoral membrane stretches between
thighs and tail but does not extend to last
tail joint; dark to light reddish brown, silky
fur; paler underparts**

DIET
**Flying insects such as flies, small moths,
beetles, mayflies and caddis flies**

BREEDING
**Breeding season: mating in fall and spring,
pregnancy in spring only (because of
delayed fertilization); number of young: 1
or 2; gestation period: 35–44 days; breeding
interval: 1 year**

LIFE SPAN
Up to 30 years

HABITAT
**Cultivated land, woodland and freshwater
habitats; roosts (and possibly hibernates)
in buildings**

DISTRIBUTION
**Throughout Europe, except Mediterranean;
further distribution not fully known**

STATUS
Very common

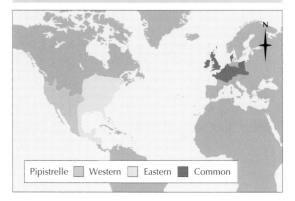

Pipistrelle ☐ Western ☐ Eastern ■ Common

The common pipistrelle lives in a wide variety of habitats, singly or in colonies, which may be small or may contain hundreds of bats. It might be found, for instance, under roofs, behind drainpipes and gutters or in holes in woodwork and brickwork, as well as in hollow trees and rock crevices. Although the common pipistrelle hibernates in the colder parts of its range, it does so only for a short period, from late October or November to March. During hibernation it will often wake up with slight temperature increases, and will then emerge to hunt for food for an hour or so. It may also move from one hibernaculum (hibernation site) to another during winter.

The American western pipistrelle bat (pictured above) is slightly smaller than the common pipistrelle, with yellow-brown fur. The former is found in the southwestern United States, from the Rockies to the Pacific.

Pictures from sound

Hunting over a regular area, the common pipistrelle flies 6–40 feet (1.8–12 m) above the ground, uttering shrill squeaks and capturing its prey on the wing. It feeds mainly on flies, particularly gnats, small beetles and moths, mayflies, caddis flies and other flying insects. It was once thought that pipistrelles continued to hunt throughout the night, but it is more likely that they have periods of activity alternating with rest periods. However, they may not return to roost until an hour or so before sunrise.

Bats use their eyes to some small extent, but mainly rely on echolocation to find their way around and to capture prey. A bat's frequent, ultrasonic squeaks bounce back from any solid object and, by interpreting the time taken for the echo to return, the bat is able to perceive its surroundings. Bats probably carry a sound-picture of familiar territory, comparable to the visual memory of human beings. Pipistrelle bats

A common pipistrelle resting on a sycamore branch. Although pipistrelle bats often mate twice in a year, pregnancy occurs only once because of delayed fertilization.

The single, occasionally two, young are born between late June and mid-July after a gestation period of 35–44 days.

The American eastern pipistrelle also has two mating periods, in late summer and early spring, but again only one litter of one to three young results, born in June or July. The females of this species usually form "maternity colonies," and the young are weaned and ready for independent flight at just 3 weeks old.

As in the European species, females of the American species are smaller than the males, which are 3½ inches (9 cm) long, including a 1½-inch (4-cm) tail. The eastern pipistrelle has a wingspan of 6 inches (15 cm) and weighs up to ¼ ounce (7 g). Its fur is yellowish brown above and paler below, with blackish wing membranes. It feeds mainly on small beetles and flies, with larger insects being held in the interfemoral membrane to be eaten. The American western pipistrelle is similar in appearance and habits but is found westward from the Rockies to the Pacific coast.

Drinking on the wing

Although pipistrelle bats eat mainly insects, the bodies of which are 70 percent water, they still need to drink. Moreover, a pipistrelle's choice for hibernation is often a cave or similar hiding place where the air is humid, sometimes so much so that the bats' bodies become beaded with moisture. This damp atmosphere prevents them from becoming dehydrated. During the summer, however, these bats' roosts, under roofs, for example, are hot and dry, and the bats must drink.

Pipistrelle bats use one of two methods to get water. Either they land on the ground, for bats are able to scuttle like mice, or they skim a pond or stream in the manner of swallows and martins, dipping their lower jaw into the water while in flight. A pipistrelle has been observed to do this several times in the course of a few minutes, swooping and maneuvering over a pond, making use of the wind like a glider pilot.

Bats can easily float on water and have no trouble flapping their way to the side of a pond or stream if they fall in. Small bats can also take off from water surfaces. However, the sides of water barrels may prove a problem, and pipistrelle bats are often found drowned in these.

appear, however, to be unable to assess size accurately. This is shown by the fact that a pebble tossed up into the path of a hunting pipistrelle will be followed down in an attempt at capture, regardless of whether the pebble is larger or smaller than the bat.

Delayed conception

Common pipistrelles are known to mate both in spring and in the fall, shortly before going into hibernation. However, because fertilization is delayed, pregnancy occurs only in the spring.

PIPIT

THE PIPITS ARE A GROUP of about 50 inconspicuous birds that, together with the wagtails, make up the family Motacillidae. Although they are usually the same brown color as many house sparrows, *Passer domesticus*, pipits have an elaborate patterning of light and dark brown on the back, head and breast. Both in appearance and in habits pipits are more like larks. They live on the ground and some sing in flight. They are small, slender birds that grow up to 6⅓ inches (16 cm), with pointed wings and fairly long legs.

The plumage of the different European species is so similar that it is quite difficult to identify pipits in the field. Identification is usually possible only by the "jizz" of a bird. This is a term used by bird-watchers to describe the individual collection of characteristics such as body shape and size, the proportions of different parts and the coloring that each species possesses.

Pipits are found on all six continents, including Antarctica. The species *Anthus antarcticus* lives on South Georgia, a large island lying within the Antarctic Convergence. Several pipits range over large areas of the Northern Hemisphere, some migrating as far south as southern Africa and Australia. The red-throated pipit, *A. vervinus*, breeds from western Alaska across northern Siberia and northern Russia to northern Fennoscandia (Norway, Sweden and Finland). It winters mostly in tropical Africa and Southeast Asia. The meadow pipit, *A. pratensis*, is found in most of Europe except in the south, and spreads into northwestern Asia, Iceland and Greenland. The tree pipit, *A. trivialis*, extends from Europe across Central Asia, and the rock pipit, *A. spinoletta*, is scattered through Europe, Asia and North America. The only other North American pipit is Sprague's pipit, *A. spraguei*, of the prairies. Six pipit species are found in South America and at least 11 live in Africa. The Bogota pipit, *A. bogotensis*, lives up to about 13,200 feet (4,000 m) in the Andes of Ecuador. Some of the African species are brightly colored. The yellow-breasted pipit, *A. chloris*, has yellow underparts, and the golden pipit, *Tmetothylacus tenellus*, is yellow all over.

Native to Africa there are eight species from the genus *Macronyx*, collectively known as longclaws, that closely resemble the *Anthus* pipits.

The yellow-throated longclaw, *M. croceus*, grows to about 7 inches (17.5 cm) long and has an exceptionally long hind claw of 2 inches (5 cm) The hind claw of other pipits is often long but is not as exaggerated as this.

Song flights

The tree pipit can often be seen perched right on the top of a tall tree, before it launches into its song flight, flying up almost vertically. Just before it levels out, it starts to sing, gliding down with its wings above its head and its tail spread. The song is more elaborate than that of the meadow or rock pipit. It consists of a single phrase of two to four notes, repeated several times, and ends with a shrill trill. The soaring flight and repetitive song that carries well is very larklike, but compared with the skylark, *Alauda arvensis*, both the flight and the song are very short. Pipits live mainly in open country such as grasslands, pampas, savanna and moorland, although some live in swamps, beside streams or along the seashore. They run rather than hop, and a few species, such as the tree pipit, have the tail-pumping habit that is characteristic of wagtails. Some pipits perch on rocks and trees. Pipits are mainly insectivorous birds.

Pipits are rather drab birds that are related to wagtails. They mainly eat ants, grasshoppers, beetles and mosquitoes, but also take slugs, spiders and a few small seeds.

The rock pipit inhabits rocky seashores and valleys up to 12,000 feet (3,600 m) in the Rockies, where it is known as the water pipit. In coastal regions it eats periwinkles and fish remains that it finds along the shore.

Nests are built in depressions in the ground, where they are often completely hidden by grass. The favorite nesting sites are in the sides of banks, cliffs and cuttings. Grasses, bents (reed-like grasses), seaweed and similar material are woven into a cup among grass and herbage. The eggs usually number four to six and are incubated by the female for about 2 weeks. The male stays nearby, feeding the female and helping to feed the chicks, which leave the nest a little after 2 weeks. Two broods are often raised in one season.

Parallel evolution

The longclaws of Africa bear a striking resemblance to the American meadowlarks, relatives of mockingbirds and grackles. They are similar in both appearance and habits, their songs are similar and their nests are built in the same way. Furthermore, when approached, they both turn away to hide their colorful underparts. This is a striking example of parallel evolution, in which two unrelated animals have evolved similar features and similar ways of life in response to the same type of environment.

The species involved are separated geographically in parallel evolution. If they lived in the same place, they would compete for food and space. Conversely, when two animals with very similar habits do live side by side, they usually exhibit slightly different behavior, perhaps in mating, feeding or choice of habitat. For example, one animal may prefer wet ground while the other favors dry ground. Such differences are sufficient to prevent interbreeding. Very similar species with slightly different habits are called sibling species. The meadow pipit and the tree pipit are sibling species: they are superficially similar and breed in the same places but in slightly different habitats. The tree pipit, for instance, starts its song flight from a tree, the meadow pipit from the ground.

AMERICAN PIPIT

CLASS	**Aves**
ORDER	**Passeriformes**
FAMILY	**Motacillidae**
GENUS AND SPECIES	***Anthus rubescens***

WEIGHT
About ¾ oz. (21 g)

LENGTH
Head to tail: 6⅓ in. (16 cm); wingspan: about 9¾–10⅔ in. (25–27 cm)

DISTINCTIVE FEATURES
Small, sleek body; slender black bill; grayish upperparts, streaked with black; white throat; peachy wash to black, streaked underparts; buff supercilium (stripe above eye)

DIET
Mainly flying insects; also some plant material in fall and winter

BREEDING
Age at first breeding: 1 year; breeding season: March–July; number of eggs: 4 to 6; incubation period: 14–15 days; fledging period: 15 days; breeding interval: 1 year

LIFE SPAN
Up to about 9 years

HABITAT
Summer: rocky tundra and birch and dwarf shrub thickets in north of range; alpine meadows and steppe to about 13,200 ft. (4,000 m) in south of range. Winter: wide range of low-altitude open habitats.

DISTRIBUTION
Summer: Alaska; northern Canada; Rocky Mountains south almost to Mexican border; western Greenland. Winter: southern U.S. and Central America.

STATUS
Common

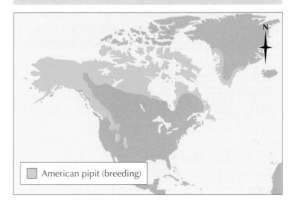

American pipit (breeding)

PIRANHA

EW ACCOUNTS OF TRAVEL in South or Central America fail to contain some references to the piranha or piraya, the small but allegedly very ferocious fish that inhabits the rivers of this region. In some places it abounds in such vast numbers as to be a serious pest, making the infested streams either very hazardous or quite impossible for people to ford or use for bathing or watering their livestock.

The name piranha applies loosely to 12 species of freshwater fish. Seven of these are members of the genus *Serrasalmus*, and have a general similarity of appearance and habits. Another three species belong to the genus *Pygocentrus*. Some scientists, however, classify these fish differently. Most of the species average about 8 inches (20 cm) in length, but the San Francisco piranha, *Pygocentrus piraya*, of the São Francisco River in eastern Brazil, reaches 2 feet (60 cm) in total length. It is one of the most dangerous of the piranha species.

Most piranhas are olive green or blue black above and silvery or dark gray on the flanks and belly. Some species have reddish or yellowish tinted fins. These colors seem to vary considerably from place to place and with age. For example, the species of the genus *Pygocentrus* might have red, orange or yellow coloration. In some species the juvenile stage is characterized by particularly vivid colors.

Piranhas have deep, short bodies that are rather compressed from side to side. A large bony crest on top of the skull supports a keel on the back, and a similar keel on the belly is strengthened by a firm row of enlarged scales. There is a fleshy, adipose fin on the back between the dorsal and tail fins. The tail is slender and muscular and together with the broad, tough, bladelike tail fin helps to drive the body through the water with great force. As in all very swift fish, the scales are very small.

Sharp teeth, muscular jaws

The most striking feature of a piranha, however, is its mouth. The massive lower jaw has relatively huge muscles and the teeth are large, flat and triangular with very sharp points. These points merely pierce the skin of prey, the actual tearing of flesh being done by the edges, which are razor-sharp. The teeth of the upper jaw are similar but much smaller and fit exactly into the spaces between the points of the lower ones

A piranha of the genus **Serrasalmus** *in the* **Pantanal, Brazil.** *Adult piranhas are voracious predators, their muscular jaws packed with razor-sharp teeth for grasping prey and tearing flesh.*

The red piranha is a popular aquarium fish. In the wild it hunts in small schools, often stalking and ambushing smaller fish.

RED PIRANHA

CLASS	**Osteichthyes**
ORDER	**Characiformes**
FAMILY	**Characidae**
GENUS AND SPECIES	***Pygocentrus nattereri***

ALTERNATIVE NAMES
Natterer's piranha; common piranha

LENGTH
Up to about 1 ft. (30 cm)

DISTINCTIVE FEATURES
Massive, muscular lower jaw; many sharp, pointed, triangular teeth; deep, short body, compressed from side to side; broad, bladelike tail fin; small scales. Adult: olive green or blue black above; silvery below; some red, orange, yellow or gold coloration. Juvenile: particularly vivid colors.

DIET
Adult: mainly insects, worms and smaller fish; sometimes attacks larger animal prey. Young: small crustaceans, fruits, seeds and aquatic plants.

BREEDING
Age at first breeding: not known; breeding season: January–February, during the rains; number of eggs: not known; hatching period: 9–10 days

LIFE SPAN
Not known

HABITAT
Fresh water, especially slow-flowing rivers, creeks and interconnected ponds

DISTRIBUTION
Rivers of central and southern South America east of the Andes, including coastal rivers of the Guianas and Brazil

STATUS
Common

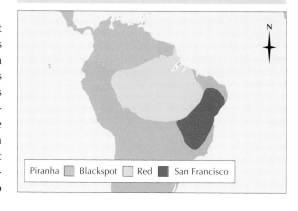

Piranha	Blackspot	Red	San Francisco

when the mouth is closed. The fact that there is a reliable record of a 100-pound (45-kg) capybara being reduced to a skeleton in less than a minute shows the efficiency of the teeth.

A few of the smaller species of piranhas are kept in aquaria, the most popular seen in tropical fish stores and public aquaria being *Pygocentrus nattereri*, the red, common or Natterer's piranha. It reaches about 1 foot (30 cm) in length.

Some piranhas are found only in certain river systems, such as the Rio São Francisco, Rio Paraguay or Rio Orinoco. The red piranha is common in creeks and interconnected ponds in Matto Grosso, Brazil, where it influences the distribution and feeding of other fish. It is also found in parts of the Rio Machado and Rio Negro. Other piranha species range over a wider area.

Stalk and ambush prey

Piranhas are more or less solitary as young, but as adults they move in groups. In some cases there is evidence of hierarchies existing within these groups. Adult piranhas are voracious predators and usually attack their prey in shoals of about 20 fish using a variety of hunting strategies including ambush, stalking or active chase to catch their prey. Smaller fish, along with insects and worms, form their staple diet, but any animal entering or falling into the water accidentally may be attacked. Piranhas are also known to attack each other. The red piranha has a highly evolved auditory capacity and exhibits a lurking, then dashing behavior when hunting. In addition, teeth replacement on alternating sides of the jaw allows continuous feeding in this species. Adults feed mainly at dusk and dawn, whereas the juveniles are active mainly during the day.

Water plant hatcheries

It is thought that piranhas breed when the rainy season sets in around January or February. The female red piranha deposits her eggs on water plants or tree roots trailing into the water. The eggs are large, adhere to plants and are not attacked by the parents. In fact, they might be guarded by the male or female, and hatch after 9 or 10 days. On hatching, the fry (young) stay attached to the vegetation in clusters until they have absorbed most of the yolk sac and then become free-swimming. The young feed on small crustaceans, fruits, seeds and aquatic plants, but once they have reached 1½ inches (4 cm) in length they begin to eat the fins and flesh of other fish that wander too close. The speckled piranha, *Serrasalmus spilopleura*, is one of the few species

that have bred in captivity. The female deposits her eggs carefully on aquatic plants, and the male guards the eggs as well as the fry when they hatch. These become free-swimming about 5 days after hatching.

Ferocity exaggerated?

The ferocity of the piranha has become almost legendary with stories of cows or pigs being stripped to a skeleton in a few minutes. A lot of these stories have been exaggerated, however, and to some extent the ferocity of the piranha is dependent on species. The red piranha and the blackspot piranha, *P. cariba*, for example, both have powerful dentition and can inflict serious bites. However, the San Francisco piranha is probably the most dangerous species and its import is prohibited in some countries. Ferocity also seems to vary from place to place, and it may be that piranhas are more aggressive at the beginning of the rainy season when the males are guarding the eggs. This could also explain why it is that they will attack bathers at certain places in a river, where perhaps they have laid eggs, leaving others unmolested not far away.

A piranha (genus Serrasalmus) among aquatic vegetation, Manu Biosphere Reserve, Peru. Many stories of the piranha's ferocity are exaggerated, but most species can inflict a serious bite.

PITTA

striking pitta is Steere's pitta, *P. steerei*, which lives in the mountain forests of the Philippines. It has a black head, a white throat, sky blue, black and scarlet underparts and a green and blue sheen on the back and wings.

Pittas live in forests, jungle or tropical scrub. There are 14 pitta species in Southeast Asia, but others are found as far east as Japan and Australia, where the noisy pitta, *P. versicolor*, is known as the dragon bird because of its gaudy colors. Two other pitta species live in Africa. Gurney's pitta, *P. gurneyi*, of south Tenasserim, southern Myanmar (Burma), is one of the rarest birds on Earth. Despite conservation efforts, it may become extinct within a few years.

Heard not seen

Despite their brilliant colors, pittas are not easy to detect in their forest habitats, as they are secretive and shy. They might be overlooked altogether if it were not for their distinctive calls, which consist of loud two- or three-note whistles, trills or grunts. When one pitta calls, others answer it.

Pittas spend most of the time on the forest floor hopping on their strong legs, but they also ascend into the trees to sing and roost. Some flick their tails continuously, as moorhens do. Several pitta species are migratory. The Indian pitta moves into Sri Lanka for the breeding season, while those pittas that breed in temperate regions, such as Japan and Australia, migrate to the Tropics. The African pitta, *P. angolensis*, breeds from central Tanzania to South Africa and parts of West Africa. Birds in the southern part of this range move north in winter to Uganda and the Central African Republic. The tropical pittas do not migrate, although they probably move about from place to place in search of fresh feeding grounds.

Pittas forage in the manner of thrushes, examining the ground with their heads cocked to one side, hopping forward to stab at an animal or to flick leaves aside and pick at anything that has been disturbed by their actions. They feed on small animals that live on the ground, including invertebrates such as earthworms, termites, beetles and centipedes as well as small lizards. Noisy pittas feed on snails in a thrushlike manner, carrying them to small stones or logs to smash open their shells.

In common with other pittas, the banded pitta, P. guajana, spends much of its time foraging in the leaf litter of the forest floor. Most species are brilliantly colored.

THE NAME PITTA IS a latinization of a word simply meaning "bird" that originated in South Madras, India. There are approximately 32 pitta species, all about the size of a thrush, with long legs, large feet and very short tails, giving them a very front-heavy appearance. They are also known as painted thrushes, ground thrushes and jewel thrushes because of their similarity to true thrushes. The plumage of pittas is immensely rich and varied, consisting of broad patches of bright colors, including greens, reds, yellows and blues.

Some species of pittas are among the most colorful birds in the world. The Indian, Bengal or blue-winged pitta, *Pitta brachyura*, which grows to about 7 inches (17.5 cm) long, is known in India as the "nine-colored one." The head is reddish yellow with a thin black band over the top of the head and broad bands running through the eyes. The back and wings are mainly green, and the rump and shoulders are pale blue. The chin and throat are white, and the rest of the underparts are reddish yellow except for a scarlet patch under the tail. Perhaps the most

AFRICAN PITTA

CLASS	**Aves**
ORDER	**Passeriformes**
FAMILY	**Pittidae**
GENUS AND SPECIES	***Pitta angolensis***

WEIGHT
About 3 oz. (84 g)

LENGTH
Head to tail: about 7 in. (18 cm)

DISTINCTIVE FEATURES
Compact body with upright stance and very short tail, giving tailless appearance; long legs and large feet; stout bill; black mask around eye; dark brown crown; white throat; green, azure and black upperparts; orange-buff breast; scarlet belly

DIET
Invertebrates, including ants, termites, bee larvae, snails and earthworms

BREEDING
Age at first breeding: 1 year; breeding season: eggs laid September–October (West Africa), November–February (East Africa); number of eggs: 1 to 3; incubation period: not known; fledging period: not known; breeding interval: 1 year

LIFE SPAN
Not known

HABITAT
Dense deciduous thickets with tangled understory

DISTRIBUTION
West Africa: resident from Guinea east through Nigeria and Cameroon to Angola. East Africa: breeds from southeast Tanzania south to eastern Zimbabwe; winters north to Central African Republic, Uganda and Kenya.

STATUS
Seldom seen; probably not rare

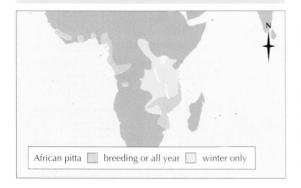

African pitta ▢ breeding or all year ▢ winter only

Pugnacious pittas

Pittas lead solitary lives. They do not tolerate any other pittas in their territories and become very aggressive during the breeding season. Some pittas, such as the African pitta, have a short display flight. The males call when they display, leaping into the air from their display post, then dropping gently back to their perch with a few shallow wingbeats. Between bouts of calling the birds sway the hips from the vertical to the horizontal while puffing out the breast feathers to expose the red belly. Calling sessions take place mainly in the early morning and evening. When the Indian pitta sees a rival, it flashes its wings and tail, showing off its brilliant colors, especially the white patches on the wings, which are normally hidden when the bird is perched on a branch.

Pittas breed when food is most abundant, which can mean virtually any month of the year in tropical Asia, depending on location, and from September to February in Africa. The nest is a domed oval mass of twigs and leaves. Its entrance is in the side and leads along a short tunnel to the nest chamber, which is lined with leaves and rootlets. The nest is usually located in a bush near the ground but may be as high as 30 feet (9 m) up a tree. The eggs are round and glossy, and in India and the Malay Peninsula and Archipelago a clutch of usually two to five eggs are laid, with fewer in Australia and South Africa. Both parents take part in incubating the eggs and feeding the young.

The Indian pitta sets up choruses of whistlings, particularly at sunrise. For this reason, it is known in Sri Lanka as the "6 o'clock bird."

PIT VIPER

An eyelash pit viper, Bothriechis schlegelii, lurks in a heliconia in Costa Rica.

as well as the Asian pit vipers, paying special attention to the organs that give them what has been called their "sixth sense," the two pits on the head that give these snakes their name.

Fold-away fangs

Pit vipers are solenoglyph snakes (see also "night adder"). That is, they have fangs that fold back and are erected only when they are about to be used. Most pit vipers are land-living, some are tree-dwellers, a few have taken to fresh water and others lead a partially burrowing life.

The water moccasin is a heavy-bodied snake that grows to 5 feet (1.5 m) in length, although most individuals are smaller than this. It mostly lives on land but hunts in the water and also takes to water when disturbed. It is black, brown or olive in color, often with rather indistinct irregular transverse bands of a lighter hue. The water moccasin is found from Virginia to Oklahoma and southward to Texas and southern Florida. Superficially it can look very like harmless water snakes, and so any semiaquatic snake within this range should be treated with extreme caution. The warning posture of the water moccasin, mouth open showing its white lining, is the origin of its alternative name of cottonmouth. While holding its mouth open, the water moccasin also vibrates its tail, like its close relatives the rattlesnakes, although it has no rattle to make a warning sound.

The copperhead is in many ways similar to the water moccasin, but the bands of color on the back are much more obvious. The copperhead grows to a total length of 4½ feet (1.4 m) and is

SOME OF THE MOST feared snakes are to be found among the 145 species of pit vipers (family Crotalidae), including well-known forms like the fer-de-lance (*Bothrops atrox*), the sidewinder (*Crotalus cerastes*) and other rattlesnakes (genera *Crotalus* and *Sistrurus*), all of which are covered elsewhere in this encyclopedia. This entry discusses other members of the Crotalidae, such as the copperhead (*Agkistrodon contortrix*), water moccasin (*A. piscivorus*) and bushmaster (*Lachesis muta*) from North America,

just as venomous as the water moccasin. It has a wider distribution than the water moccasin, ranging from Massachusetts and Illinois to western Texas, but it is not found in central and southern Florida. The largest pit viper in the Americas is the bushmaster, which may grow to 12 feet (3.7 m) in length. The bushmaster is found from Costa Rica to Colombia, and closely related species occur as far south as Argentina. The bushmaster and its relatives are among the most feared of the world's venomous snakes.

PIT VIPERS

CLASS	**Reptilia**
ORDER	**Squamata**
SUBORDER	**Serpentes**
FAMILY	**Crotalidae**
GENUS	**Several, including *Agkistrodon*, *Bothrops*, *Lachesis* and *Trimeresurus***
SPECIES	**About 150, including copperhead, *Agkistrodon contortrix*; Himalayan pit viper, *A. himalayanus*; water moccasin, *A. piscivorus*; bushmaster, *Lachesis muta*; habu, *Trimeresurus flavoviridis*; and Wagler's pit viper, *T. wagleri***

ALTERNATIVE NAMES
Cottonmouth, gapper, trapjaw (water moccasin); highland moccasin, chunkhead (copperhead); Japanese pit viper (habu)

LENGTH
Water moccasin and copperhead: up to 4 ft. (1.2 m); bushmaster: up to 12 ft. (3.7 m)

DISTINCTIVE FEATURES
Heat sensitive organs located on either side of head between nostril and eye; all species highly venomous

DIET
Mainly mammals and birds

BREEDING
Most species: ovoviviparous (live young); bushmaster and relatives: oviparous (lay eggs)

LIFE SPAN
Not known

HABITAT
Very varied

DISTRIBUTION
North, Central and South America; western, central and Southeast Asia; Japanese islands

STATUS
Some species abundant

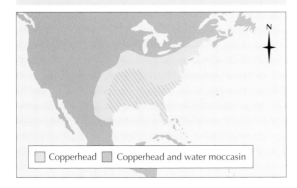

Copperhead ◻ Copperhead and water moccasin

An eyelash pit viper performing a threat display. Such a display is what got its North American cousin, the water moccasin, its alternative name of cottonmouth.

The Asian pit vipers are of two kinds, tree-dwelling and ground-living, the first having prehensile tails that assist their climbing. The Himalayan pit viper, *Agkistrodon himalayanus*, lives at 7,000–16,000 feet (2,100–4,900 m), sometimes being found even at the feet of glaciers. The Asian pit vipers are found mainly in eastern Asia and Southeast Asia, with one species ranging as far west as the mouth of the Volga River. Wagler's pit viper, *Tremeresurus wagleri*, is kept in large numbers in the Snake Temple in Penang. The most feared snake in Japan, the habu, *T. flavoviridis*, is a pit viper.

Warm-blooded food

Pit vipers eat a wide range of foods. The water moccasin, for example, eats rabbits, muskrats, ducks, fish, frogs, other snakes, birds' eggs and nestlings. The copperhead eats small rodents, especially mice, other snakes, frogs, toads and insects, including caterpillars and cicadas. The bushmaster, by contrast, eats mainly mammals, and pit vipers generally tend to hunt warm-blooded animals more than cold-blooded species, as one would expect from snakes with heat-detector pits. Pit vipers have one pit on each side of the head between the eye and the nostril. Using these, a pit viper can pick up the trail of a warm-blooded creature.

understandable terms, a pit viper could detect the warmth of a human hand held 1 foot (30 cm) from its head. The membrane with its receptors can be compared to an eye with its retina. The overhanging lip of the pit casts "heat shadows" onto the membrane, so the snake is aware of direction, and since the "fields of view" of the two pits overlap, there is the equivalent of stereoscopic vision, enabling the snake to determine range. A pit viper hunting by day can use its eyesight to track an animal, employing scent to follow its trail through low vegetation after the animal has passed out of sight, and with its pits the snake has the advantage of being able to perceive its prey's heat signature as well. However, it is in night hunting that the facial pits really come into their own, when prey can be tracked by scent, with the facial pits, rather than eyesight, guiding the final strike. At first it was thought the pits had something to do with an accessory aid to smell or as an organ of hearing (snakes have no ears). Another suggestion was that they might be organs for picking up low-frequency air vibrations. Then in 1892 it was noticed that a rattlesnake, one of the pit vipers, was attracted to a lighted match. After that came the discovery that pythons have pits on their lips that are sensitive to heat. The first experiments on pit vipers were carried out in 1937 and left no doubt that the pits are heat detectors. Further studies since have shown just how delicate they are.

Snakes in cold climates

Pit vipers usually bear live young. There are exceptions, the bushmaster being one. One of the advantages of bearing live young, as against laying eggs, is that the offspring are protected not only from predators but also against low temperatures until they are at an advanced stage of development. At

This white-lipped pit viper, Trimeresurus albolabris, *provides a good view of the pit situated between its eye and its nostril. Most snakes of this Asian genus of pit vipers, especially green ones, live in trees.*

"Seeing" heat

Each pit is ⅛ inch (3 mm) across and ¼ inch (6 mm) deep. A thin membrane is stretched near the bottom, and temperature receptors are packed within this membrane at a density of up to 1 million per square inch (1,500 per sq mm). These receptors are so sensitive they can respond to changes as small as 0.0036° F (0.002° C) and allow a snake to locate objects 0.18° F (0.1° C) warmer or cooler than the surroundings. In more

some time pit vipers must have crossed the land bridge that used to exist where the Bering Strait is now. This is a long way north, and it would have been far easier for snakes able to bear live young to survive in these latitudes and so make the crossing. It probably also explains why the Himalayan pit viper can live so near glaciers and why the most southerly of all snakes is a pit viper named *Bothrops ammodytoides*, which lives in the Santa Cruz province of Argentina.

PLAICE

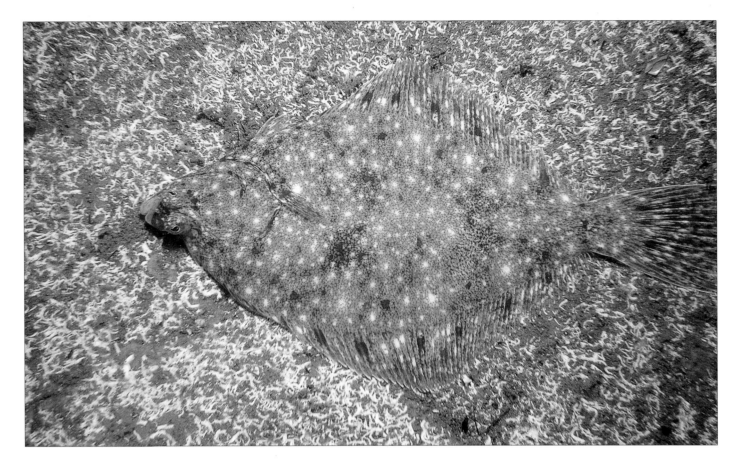

T HE PLAICE IS ONE OF THE best known of the flatfish and commercially the most important. It has a flattened body, with the dorsal fin extending from the head almost to the tail fin, and the anal fin runs from behind the gill cover to the same point.

The fish's brownish upperside is marked with red spots, each of which is surrounded by a white ring in the adult. These may be pale when the fish has been resting on whitish pebbles. The underside is pearly white but can be partially or wholly colored, a feature known as ambicoloration. It may take the form of scattered brown or black spots or patches on the white undersurface. Alternatively, only the hind end may be completely colored as on the upper surface, including the red spots. When the pigmentation extends along the whole underside, the undersurface of the head is usually white, but in exceptional cases even this may be colored. The mouth is twisted, with the lower, or blind, side more developed and armed with a greater number of teeth. The small scales are embedded in the skin, and there are bony knobs between the eyes. Plaice can grow to almost 39 inches (1 m) long, although most are usually much smaller than this.

Plaice range from Iceland and the White Sea (the waters to the north of European Russia), along the coasts of Scandinavia, south through the North Sea to the coasts of the Mediterranean. Plaice are not identical throughout their range but are split into a number of subspecies. They vary in area of distribution, time and location of spawning and in their degree of pigmentation.

Life at the bottom

Plaice live on sandy, gravelly or muddy bottoms, alternating between remaining slightly buried and swimming just off the bottom at intervals through the day and night. They swim with vertical undulations of the flattened body. When they want to descend to the bottom, plaice hold the body rigid and glide downward. On touching the bottom, they undulate the fins to disturb sand or mud, which then settles on the fins, disguising the outline of the body. In this position plaice breathe with a suction pump action of the gill covers.

Young plaice seem to go into a state resembling hibernation during the winter. They remain quiescent in shallow water, slightly buried in the sand, moving from the shallow water to deeper water as they mature.

A plaice can change its color and patterning to blend in with its background. A hormone secreted by the fish alters the shape of the skin's pigment cells, dramatically changing the fish's appearance.

The characteristic orange spots on the plaice's upperside become especially vivid during the spawning season.

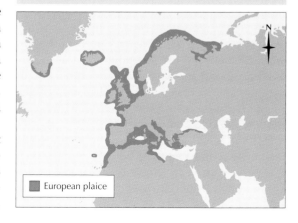

EUROPEAN PLAICE

CLASS **Osteichthyes**

ORDER **Pleuronectiformes**

FAMILY **Pleuronectidae**

GENUS AND SPECIES ***Pleuronectes platessus***

WEIGHT
Up to 15½ lb. (7 kg)

LENGTH
Up to 39 in. (1 m), usually much smaller

DISTINCTIVE FEATURES
Flattened body; both eyes on right side of head; relatively small head and jaws; twisted, asymmetrical mouth; "eyed" side brown with large bright red or orange spots; "blind" side clear or pearly white

DIET
Mainly thin-shelled mollusks and polychaetes (marine worms)

BREEDING
Age at first breeding: 2–3 years (male), 4–5 years (female); breeding season: January–May (northern North Sea), October–March (southern North Sea); number of eggs: 50,000 to 400,000; hatching period: 10–20 days, depending on temperature; breeding interval: 1 year

LIFE SPAN
Up to 50 years

HABITAT
Coastal waters on sand, gravel or mud seabeds; older fish in deeper waters

DISTRIBUTION
Southern Greenland, Iceland and White Sea (Arctic waters to north of European Russia), south to Mediterranean and Morocco

STATUS
Common

European plaice

Chisel and grinder

The teeth in the jaws of plaice are chisel-like, but the pharyngeal (throat) teeth are blunt and act as crushing tools. The fish feed mainly on small mollusks, but also take other small bottom-living invertebrates, such as worms.

Plaice swim over the shore at high tide to feed on cockle and mussel beds. They hunt by sight, not raising the head far off the bottom but shooting forward horizontally with great accuracy to take the prey. Very small mollusks are taken whole into the stomach. Larger ones are crushed by the pharyngeal teeth. Plaice also bite off the siphons of mollusks or the heads of worms sticking out of tubes.

Prolific spawnings

There is little in the outward appearance of plaice to distinguish the male from the female. However, if they are held up to the light at spawning time, a small, dark triangle is visible on the female roe. The male roe is a curved rounded line. The male plaice reach the spawning grounds first and remain there until after the females have gone. Spawning time differs from one part of the sea to another. Off the eastern coast of Scotland spawning takes place from early January–May, with a peak in March. In the Clyde estuary, on the western coast of Scotland, it lasts from February–June. In the southern North Sea it takes place from October–March.

To spawn, two plaice swim about 2½ feet (75 cm) off the bottom, the female lying diagonally across the male, releasing a stream of eggs while he emits a stream of milt (sperm-containing fluid). Spawning lasts less than a minute, after which the two separate and return to the bottom. Each female lays 50,000 to 400,000 eggs; the precise number of eggs seems to depend on the length of the fish. The transparent eggs, each in a tough capsule, are slightly more than 2 millimeters in diameter. They float at or near the surface, and many are eaten before they can hatch, which occurs in 10–20 days, according to the temperature of the water. Unlike the eggs of other fish, plaice eggs do not contain an oil globule to help them float the surface. They require a high level of salt in the water in order to remain buoyant. For this reason, spawning occurs in water with high levels of salinity. The larvae are about ¼ inch (0.6 cm) long, without mouth or gills, and with the remains of a yolk sac attached, which nourishes them.

This period is the most vulnerable time in the life of a plaice. Apart from those plaice that are eaten by other animals, only 1 in every 100,000 survive the first few weeks of larval life, or 2 to 5 for every pair of parent plaice. Although this suggests a low survival rate, in one area alone, halfway between the mouth of the Thames River on the southeast coast of England and the coast of Holland, 60 million plaice come together each year to spawn. The adults are probably protected from attack by their color and their habit of lying buried. However, seals find them, and predatory fish, such as cod, eat the small plaice.

Juvenile diet

As the contents of the yolk sac are gradually used up, the larval plaice starts to feed on diatoms (planktonic algae). At this stage it has the normal fish larva shape, and has none of the physical characteristics of the adult plaice. As it grows, the young fish graduates from eating small diatoms to feeding on larger diatoms and then to larvae of small crustaceans, such as copepods, and mollusks. At this stage in the plaice's development, it feeds heavily on the planktonic food *Oikopleura*. After 2 months the larva gradually metamorphoses into a young flatfish; this takes about 2½ weeks. The body becomes flattened from side to side, the young plaice starts swimming on its side and the skull becomes twisted by growing more quickly on one side than the other, causing the left eye to migrate to the right side. At the same time the young plaice leaves the upper waters for the seabed, settling on its left side so its right side and both eyes are uppermost. While these changes are taking place, the young plaice, which by this stage is about ½ inch (1.25 cm) long, is transported by currents to its inshore nursery ground.

The account given above regarding the feeding patterns of plaice larvae is necessarily a generalization. The food taken varies in different places, and the larvae take whatever food is available. In Scottish coastal waters they eat mainly worm larvae, crustacean eggs and larval mollusks. Off Plymouth, on the southwest coast of England, copepods and other small crustaceans are eaten. In the Irish Sea the larvae feed on small copepods and spores of algae, while in the southern North Sea they feed mainly on *Oikopleura*. The survival of the larvae can be seriously affected if supplies of these foods are low in a given area.

Development of the young

After the ½-inch (1.25-cm) young plaice has settled on the bottom its growth rate accelerates. The female plaice reaches 2 inches (5 cm) by the age of 1 year, 5 inches (12.5 cm) by 2 years, nearly 8 inches (20 cm) by 3 years, 10½ inches (27 cm) by 4 years and 13 inches (33 cm) by 5 years of age. The male is smaller than the female and, on average, reaches sexual maturity in 2–3 years, whereas the female matures in 4–5 years. However, these figures are approximations because the average size of plaice has been found to vary considerably according to environmental conditions. Some specimens found in the North Sea measure 17 inches (43 cm), while others in the English Channel average 15 inches (38 cm). Specimens found in the Kattegat measure 13 inches (33 cm) and in the Baltic plaice average 10 inches (25 cm). After 20 years plaice generally measure about 2 feet (60 cm) and a 33-inch (84-cm) plaice would be about 40 years old.

The plaice has a twisted, asymmetrical mouth that it uses to suck crustaceans and mussels onto its pharyngeal teeth.

PLAINS-WANDERER

In a family of its own, the plains-wanderer is distinguishable from its near relatives the button quails both by its hind toe and by its upright posture.

ALSO KNOWN AS TURKEY QUAILS, plains-wanderers are unusual and now rare Australian birds that are the sole members of the family Pedionomidae. Along with their relatives the button quails, and with painted snipe and phalaropes (all of which are discussed elsewhere in this encyclopedia), they share the unusual feature of having dominant females. The female plains-wanderer grows up to 7½ inches (19 cm) long, considerably longer than the male, and has much brighter plumage. The plumage is not gaudy but has intricate patterns of reddish brown, buff and black. The wings are short and rounded. At one time the plains-wanderer was classified with the button quails, to which it is very similar in appearance,

although it has certain structural differences. It has a hind toe and eggs that are pointed rather than oval.

Plains-wanderers live in south-eastern Australia, from eastern parts of the state of South Australia, north to Queensland and east to Victoria and New South Wales. They seem to be most abundant in northern Victoria and New South Wales and keep to the dry plains and grasslands, avoiding the scrub regions where the button quails live. The plains-wanderers' preferred habitat of sparse lowland native grasslands is one of the most depleted and threatened habitats in southern Australia. To suit the birds' requirements the land must be treeless and have bare patches; ideally, it should be lightly grazed as well. Plains-wanderers also favor the regenerating stubble of cereal fields. They feed mainly on seeds from grasses and other low-growing plants, together with some insects.

Reluctant fliers

Plains-wanderers are usually active at night and move about in pairs or family parties rather than in flocks as their relatives do. Although they are not a migratory species, plains-wanderers do move from place to place in search of new feeding grounds. On such nomadic excursions, they fly with rapidly whirring wings, although they are generally very unwilling to fly. If they are disturbed, the birds run through the grass, occasionally rising on tiptoe to peer around, or freeze, crouching motionless. It is sometimes possible to catch them by hand. Observers have described the plains-wanderers' style of running as being remarkably ratlike.

Female initiates courtship

The breeding habits of the plains-wanderer have not been studied in detail, due in part to the species' rarity. However, it is known that the female takes an active part in courtship and that the male incubates the eggs and guards the chicks. Scientists are not sure whether the female mates with several males, as in button quails.

The nest is a grass-lined depression usually sheltered by tall grass or a bush, which acts both as a shield from predators and as protection from

PLAINS-WANDERER

CLASS **Aves**

ORDER **Gruiformes**

FAMILY **Pedionomidae**

GENUS AND SPECIES **_Pedionomus torquatus_**

ALTERNATIVE NAMES
Collared plains-wanderer; turkey quail

LENGTH
**Head to tail: male, about 6 in. (15 cm);
female, 6¾–7½ in. (17–19 cm)**

DISTINCTIVE FEATURES
**Resembles large button quail. Compact body;
slender neck; yellow bill; pale eyes; fairly
long, yellow legs with hind toe on each foot.
Female: intricate patterns of buff, brown,
black and reddish orange; rufous breast band;
white collar with black spots. Male: much
paler and plainer; no breast band or collar.**

DIET
**Mainly seeds of grasses and other
low-growing plants; some invertebrates**

BREEDING
**Age at first breeding: 1 year; breeding
season: fall and early winter (Queensland),
early spring to summer (rest of range);
number of eggs: usually 4; incubation
period: about 24 days; breeding interval:
varies according to rainfall**

LIFE SPAN
Not known

HABITAT
**Sparse, treeless, lightly grazed grasslands
with areas of bare ground; also fields of
cereal stubble and sparse saltbush**

DISTRIBUTION
**Highly localized range in eastern and southern
Australia, particularly in New South Wales**

STATUS
**Vulnerable; estimated population: no more
than 8,000 to 11,000**

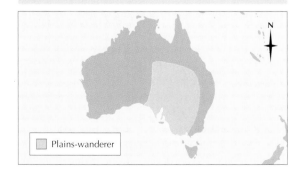

Plains-wanderer

the midday sun. There are four eggs to a clutch, pale yellow or green with gray and brown spots. The male incubates the clutch and rears the young without the female, and the incubation period is about 3½ weeks. The chicks leave the nest soon after hatching.

Plains-wanderers communicate with a repeated low _moo_, sometimes described as _oom oom oom_ or _whoo whoo whoo_. They also make soft clucking sounds.

A rare photograph of an obscure and highly localized species. The usual roles of the sexes are reversed in plains-wanderers. The female (above) is brighter than the male, and it is the male that takes care of incubation.

Causes of decline

The plains-wanderer is one of the growing number of animals that are becoming rare or extinct before their habits have been studied. In recent times many birds have become extinct before their eggs and incubation periods have been described. At one time the plains-wanderer was quite common, but after Europeans settled in Australia its life became more and more precarious. The grasslands have been transformed by agriculture; even stock farming drives out the plains-wanderer because sheep eat the seed-bearing plants on which it lives. Introduced rabbits and fires have had the same effect, and those that survive fall prey to foxes and feral cats. The World Conservation Union (I.U.C.N.) classes the plains-wanderer as vulnerable and the bird is now extinct in much of its range. Population estimates for the species vary, due to the difficulty of counting this secretive, nocturnal, highly nomadic bird. However, there are probably 8,000 to 11,000 birds left in the wild.

PLANKTON

Plankton sometimes drift through the water in sizable colonies. The colony of planktonic ascidians (sea squirts) shown above is about 4 feet (1.2 m) long.

PLANKTON IS A WORD OF Greek derivation meaning literally "that which drifts." For this reason the term is applied to aerial, freshwater or saltwater drifting organisms. This article will consider the marine forms of plankton, which are by far the most spectacular and colorful of all drifters.

Marine plankton drift with oceanic currents, rapid shoreline undertows and windblown surface currents, but they are far from inactive. Indeed, perhaps the most noticeable feature of a zooplankton (animal plankton) sample is that every animal quivers, jerks and undulates in constant activity. Under the microscope the swimming movements often appear efficient and fast. Nonetheless, the progress, when compared with the forces and speeds of oceanic currents, is very limited. Some of the more capable swimmers do, however, manage to migrate vertically, and so pass from one current of water to another. By dropping from, for example, an east-going current into a lower west-going current, the tiny animal can ride the currents to maintain its position over the sea floor or control its drift.

Community members

Drifting animals vary considerably in size and may be found at any depth in the oceans, but the majority are confined to the upper 100 feet (30 m). Plankton communities comprise both animals and plants, and these small, highly specialized drifters feed on plant growth or on each other in the same way that land animals do. In general, the phytoplankton (plant members of the plankton community, also called microalgae) have no means of swimming. They are eaten by herbivorous animals or zooplankton. These plant-eaters, in turn, are caught and devoured by carnivorous animals. These may then be the victims of nonplanktonic, predacious marine animals, such as fish or squid or even some whales, forming a food web, or series of food chains, as well defined as those found in other animal or plant communities. Energy is derived from sunlight filtering through the surface waters. This is converted to vegetable food by phytoplankton and plants and is then passed on to zooplankton. Because their energy is derived from the sun, most of the organisms are found near the water's surface.

Most planktonic organisms are small. There are, however, a large number of jellyfish that are truly planktonic and may reach a considerable size. For instance, the common orange-brown *Cyanea* jellyfish has an Arctic relative that may be 6 feet (1.8 m) across the bell and have tentacles 18 feet (5.4 m) long. The Portuguese man-of-war, *Physalia physalis*, may have trailing tentacles

50 feet (15 m) or more in length. At the other extreme, many planktonic organisms are unicellular and are no larger than an amoeba. Indeed, there are large numbers of amoeba-like creatures in the plankton that have within their microscopic bodies tiny transparent skeletons of perfect symmetry, yet they are so small as to be invisible to the naked eye. The great majority of planktonic organisms, however, are about the size of a water flea, or between the size of a pinpoint and a pinhead. The oceans of the world teem with billions of these animals and plants.

One of the most intriguing facts about zooplankton is that many of the creatures are the minute young or larval stages of much bigger, often common, shoreline or open-water adults. One vast topic within the study of plankton is the investigation of these larval forms. In the same way that a frog begins its life as a tadpole, or a moth as a caterpillar, so a starfish begins its existence as a bipinnarian or a sea snail as a veliger. The larvae of the familiar seashore animals are usually small, highly specialized and well adapted to a drifting existence. The adults are usually adapted to a sedentary or creeping existence. Because of this extreme larval specialization, it is not surprising to find that these planktonic stages of a life cycle are often radically different in appearance to the adult forms.

Unrecognizable offspring

The young stages of plankton look so unlike their adult counterparts that they were named as separate animals when they were first studied by biologists. The planktonic larva of the European freshwater eel, *Anguilla anguilla*, was called *Leptocephalus*, but it is now known as the eel leptocephalus larva. The sea cucumber's early form was originally named *Auricularia*, whereas today it is called the auricularian larva of a sea cucumber. The Latin word *auricularia* refers to an earlike shape; *leptocephalus* describes a flat-headed appearance.

Some of the first larvae to be linked with their adult form were the zoeae larvae and megalopa or second-stage larvae of crabs. The correlation between barnacles and their larvae was recognized in the mid-19th century. These organisms begin life as nauplii, pass through six very active stages and then turn into small bivalved settling larvae that attach themselves to rocks and metamorphose into barnacles. After a time of feeding, growth and sometimes skin change, there is a sudden, brief period of more rapid change and the

larva appears as a diminutive adult. Why do these shoreline animals have planktonic larvae? Essentially, they do so in order to distribute the species far and wide. It would be problematic for barnacles to colonize rocks and boat bottoms, or for a permanently fixed sponge to disperse, without these mobile drifting larvae. Sometimes the larval stages inform a biologist about the affinities of the adult. A barnacle is known to be a crustacean because its nauplius larva, the telltale feature, is a typical crustacean larva. However, many planktonic creatures have not yet been fully studied and many larvae are not linked with adults.

Blooms and red tides

In open ocean, plankton tends to be at low density and relatively stable. Indeed, the most central parts of the major oceans, especially those in the Tropics, are often referred to as the oceanic equivalent of deserts. This is largely true of some tropical inshore areas too. However, in shallow temperate seas, the situation is very different.

In winter, lack of light results in relatively low numbers of plankton in shallow waters, but as the length of the days increases during the year, the growth rate of phytoplankton begins to increase rapidly, and they may exist in densities hundreds or thousands of times higher in late spring and summer than during winter. These very high densities of phytoplankton are venomous blooms. They are typically detectable as thick, soupy water of low clarity, but may

Seen against the eye of a needle, the seawater droplet below contains a variety of plankton, including copepods and crab larvae.

form thicker reddish, yellow or brown scums on the water's surface. They are commonly known as red tides.

Over the summer, the dense phytoplankton deplete the plant nutrients in the seawater, especially nitrogen, and phytoplankton density usually becomes greatly reduced due to death and their being taken by zooplankton. Consequently, the seawater may become very clear during this period. Species composition may change dramatically too. Spring phytoplankton blooms are often dominated by large species of diatoms (single-celled planktonic algae). In summer smaller types, such as dinoflagellates (single-celled plankton that belong to the order Dinoflagellata), are more important.

Sometimes the winds of the fall equinox stir up the seabed enough to reintroduce plant nutrients into the seawater, and a mini fall bloom may occur. As the daylight begins to decrease thereafter, plankton levels decline for the winter, during which storms disturb the seabed sediments and reintroduce nutrients into the water for the following year.

As they grow, zooplankton begin displaying adult characteristics. This shrimp larva has long, paddlelike antennae that it uses to propel itself through the water.

Toxic blooms

Very dense blooms of phytoplankton may clog fishing nets and fish gills. They may also reducing the oxygen content in the water, sometimes to the point at which fish and other animals may be killed in large numbers. Some bloom species can be very toxic. When they are filtered, for example by mussels, they become concentrated and may present a serious threat to human health if consumed. Numerous deaths have been brought about by this form of poisoning. There are various degrees and types of symptoms, depending on species, concentration and the susceptibility of the victim, but they may be generally described under the following three descriptive headings: amnesic shellfish poisoning, diuretic shellfish poisoning and paralytic shellfish poisoning.

Bizarre rarities

Some of the most visually arresting species of plankton are also the most common. One such type, a copepod (minute marine and freshwater organism), occurs in Atlantic coastal waters. It resembles a water flea, but within its body it is equipped with two striking binocular eyes. The retina is halfway back along the body and the lens system is mounted at the front of the head. Between retina and lens is a condenser lens, and the whole system is bounded by a well-defined binocular tube.

Planktonic organisms are essential to the vast food chains in the oceans. Marine life depends on them, from immense baleen whales, which filter them directly from the water, to predatory fish that eat them secondhand in the bodies of their prey. The study of the oceans' plants and animals and their interrelationships presents even the amateur biologist with an open field for research.

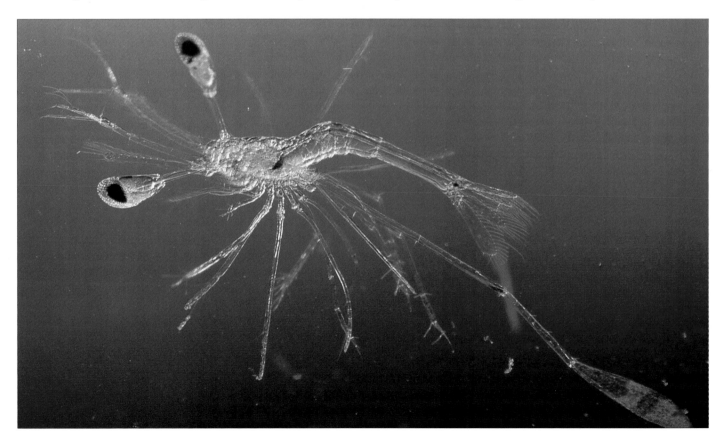

PLATY

THE PLATY IS A SMALL TROPICAL FISH, and is a long-standing favorite with aquarists, who, over time, have abbreviated its generic name, *Platypoecilus*, to platy. The genus has now been changed to *Xiphophorus* but platy, or platyfish, still persists as the common name.

The southern platy, *X. muculatus*, is a deep-bodied freshwater fish of Mexico and Guatemala with a single dorsal fin and a relatively large, rounded tail fin. The males are up to 1¾ inches (4.5 cm) long, and the females grow up to 2¼ inches (5.5 cm). The color is brownish to dark olive on the back, the flanks being bluish and the underside whitish. The fins are almost transparent, but in the male the pectoral fins are bluish at the tips and the anal and tail fins have a greenish white band. In the past, platys have also been called moonfish because of the mark at the base of the tail fin, which looks like a crescent moon. However, the name is seldom used today and the mark is variable or missing in many individuals. The colors are also variable, even in the wild forms, in which black, checkered and red varieties are not uncommon.

A second species, from another part of Mexico, was named the variable or variegated platy, *X. variatus*, because of the range of colors it shows. It is similar to the first species, and some experts treat the two as a single species.

Rainbow colors

Platys do not swim together in unison, but form loose groups to which there is a vaguely hierarchical structure, led by the more strikingly colored dominant males. Although they are a popular aquarium fish, platys have a high tolerance of salt and wild platys are often found in slow-moving brackish water. Platys are easy to keep and they breed well, often producing color varieties that can be selected and breed true. They have probably provided more color varieties than any other aquarium fish. Among these are the blue platys, also known as the blue moon or blue coral platy, which come in a wide range of blues; the red platys, which at first were red, stippled with the black dots of the wild platy but are now bred in a pure deep red form; and the black platy, which is green or yellow with a broad black stripe along the body or all black except for the fins. A yellow variety that one European aquarist found in his stock became known as the golden platy. Crossed with the wild form, it produced the wagtail platy, which initially had black fins and a gray body. After several generations of breeding, the body has become yellow.

Brief courtship

As well as eating water fleas, mosquitoes and midge larvae, platys take plant food and are particularly fond of nibbling the green algae that grow on the sides of aquariums. Courtship and mating are very brief. The colors of the male intensify during the breeding season and he either swims alongside the female or dashes about, encircling her with outspread fins. The couple mate abruptly, in one quick movement, while both partners are still moving forward. Fertilization is internal and the babies are born

The hi-fin platy is so called because of its high, trailing dorsal fin. Native to Mexico, Guatemala and Belize, platys are among the most colorful and popular aquarium fish.

SOUTHERN PLATY

CLASS	Osteichthyes
ORDER	Atheriniformes
FAMILY	Poeciliidae

GENUS AND SPECIES *Xiphophorus maculatus* (formerly *Platypoecilus maculatus*)

ALTERNATIVE NAMES
Southern platyfish; moonfish (archaic); many captive-bred varieties have names

LENGTH
Up to 2¼ in. (5.5 cm)

DISTINCTIVE FEATURES
Small size; single dorsal fin; rounded tail fin. Wild form: brownish olive nape and back; bluish flanks; whitish throat and underside; blackish blotch at base of tail fin; colorless, almost transparent fins. Coloration highly variable, in both wild and captive-bred fish; black, checkered and red varieties also occur.

DIET
Worms, crustaceans, insects, plant matter

BREEDING
Age at first breeding: 3–4 months; number of eggs: up to 100

LIFE SPAN
Up to 2 years

HABITAT
Warm springs, weedy canals and ditches

DISTRIBUTION
Mexico south to northern Belize

STATUS
Common

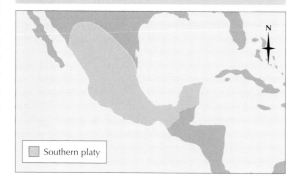

Southern platy

Platys are communal fish, moving together in small, loose shoals. The shoals have a loose hierarchical structure in which the more strikingly colored males are dominant.

alive, 10 to 75 at a time. The welfare of the young depends on the temperature of the water; for example, they grow noticeably faster and survive better at 97° F (35° C) than at 103° F (39° C).

In aquariums platys readily hybridize with swordtails, which belong to the same genus, yet although these two types of fish live almost side by side in their natural habitat, they have not been known to hybridize there. Careful study suggests that courtship is not only a way of bringing the sexes into breeding condition, but also a kind of language by which members of the species recognize each other and learn to tell the difference between their own kind and members of another species.

The courtship of the platys and swordtails is very similar, differing only in a number of small details. None of these is important on its own, but together they keep members of the different species apart. In an aquarium, however, where there is not the same wide choice of partners, the "language" of courtship breaks down and a male and a female of different species will come together and breed, giving rise to hybrids.

Platys and cancer research

A medical discovery of significant importance resulted from one hybridization between a platy and a swordtail. When a spotted platy was crossed with a green swordtail, *X. helleri*, the offspring consistently developed cancerous growths along the side of the body, remarkably like a certain type of cancer that occurs in human beings. This provided an example of a heritable

cancer, one that could be produced to order merely by crossing two kinds of fish. The discovery made it possible to study the genetics of cancer in a species that bred rapidly, giving quick results, whereas comparable studies on humans would be long and difficult.

PLATYPUS

WITH ITS DUCKLIKE BILL, furry mammalian coat and webbed feet, it is not surprising that the platypus astounded the scientific world on its discovery. So strange did the animal seem to Western eyes that one scientist named it *paradoxus*. The platypus was not discovered until 1796, nearly 200 years after the first wallaby, for instance, had been seen by a European. Its fairly recent discovery is due to the fact that aquatic animals are often elusive and may avoid detection for some time.

Unusual physical adaptations

Also known as the duckbill, watermole or duckmole, the platypus is one of Australia's two egg-laying mammals, the other being the spiny anteater or echidna. The platypus is 12–18 inches (30–45 cm) long, including a 4–6-inch (10–15-cm) beaverlike tail, and it weighs 1–4½ pounds (0.5–2 kg), the males being slightly larger than the females. The "bill" is a sensitive, elongated snout and is soft.

The platypus is well adapted to its semi-aquatic life. The legs are short with strong claws on the toes and the feet are webbed. The webbing on the forefeet extends well beyond the toes but can be turned back when the animal is on land, leaving the claws free for walking and digging. The eye and the opening to the inner ear lie on each side of the head in a furrow that can be closed when the platypus submerges. There are no external ears, so the platypus is blind and deaf when it is underwater.

Thick, loose skin makes the barrel-shaped body of the platypus appear larger than it is. The pelt consists of a dense, woolly undercoat and long, shiny guard hairs. The color varies from sepia brown to almost black above and is silver, tinged with pink or yellow underneath. Females can be identified by the more pronounced reddish tint of their fur.

Adult males have hollow spurs on the ankle of each hind limb, connected to venom glands in the thigh. These spurs are usually covered by a fold of skin. The poison is a transparent liquid and kills its victim by thickening the blood. It can be quite harmful to humans, and can kill dogs. The venom becomes more toxic during the breeding season and the poison glands also grow larger during this time. Male platypuses use their poison spurs to fight each other during the breeding season. However, it is likely that the poison spur evolved to combat predators that have since become extinct, rather than to kill rivals of the same species.

The western limits of the platypus's range are the Leichhardt River in North Queensland, and the Murray, Onkaparinga and Glenelg

When underwater the platypus uses its strong forefeet for swimming and its hind legs as rudders.

PLATYPUS

CLASS	**Mammalia**
ORDER	**Monotremata**
FAMILY	**Ornithorhynchidae**
GENUS AND SPECIES	***Ornithorhynchus anatinus***

ALTERNATIVE NAMES
Watermole; duckmole; duckbill

WEIGHT
1–4½ lb. (0.5–2 kg)

LENGTH
**Head and body: 12–18 in. (30–45 cm);
tail: 4–6 in. (10–15 cm)**

DISTINCTIVE FEATURES
**Streamlined, flattish body; soft, elongated
snout resembles a duck bill; short legs;
webbed feet with strong claws; beaverlike
tail; red-brown above; lighter gray or silver
beneath; female more red than male**

DIET
**Freshwater shrimps, crayfish, insect larvae,
worms, tadpoles and small fish**

BREEDING
**Age at first breeding: 2–3 years; breeding
season: July–November; number of eggs: 1
to 3; gestation period: eggs laid about 27
days after mating; hatching period: 10–12
days; breeding interval: 1–2 years**

LIFE SPAN
Up to 13 years

HABITAT
Streams, rivers, ponds and lakes

DISTRIBUTION
Eastern and southern Australia; Tasmania

STATUS
**Locally common, but vulnerable to habitat
disturbance, fish netting and river pollution**

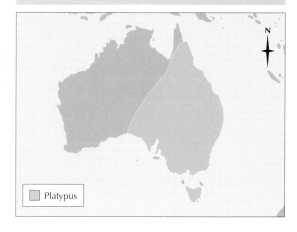

Platypus

---

*The platypus's webbed
feet provide it with
considerable power
and maneuverability
underwater. Each foot
has five clawed toes
used for digging.*

Rivers, just within the border of South Australia.
It is found in all forms of fresh water, from clear
icy streams at 5,000 feet (1,500 m) to lakes and
warm coastal rivers.

Voracious feeder

In common with many small, energetic animals,
the platypus has a voracious appetite and probably needs more food, relative to its weight, than
any other mammal. It feeds mainly in the early
morning and late evening on crayfish, worms
and other small water animals. It probes for these
with its bill and at the same time takes in grit and
sand, which probably help the animal break up
its food. Young platypuses have teeth, but these
are replaced in the adult by horny ridges. By day
the platypus rests in burrows dug out of the
banks, emerging in the evening to forage for food
in the mud of the river bottom.

An egg-laying mammal

The platypus's breeding season lasts from July to
November, and mating takes place in the water
after an elaborate and unusual courtship. Among
other maneuvers, the male grasps the female's
tail, and the two then swim slowly in circles.

During this period the male uses a simple
burrow in a riverbank. The female later digs a
nursery. Winding and intricate, this much larger
burrow is usually 25–35 feet (7.5–10.5 m) long,
although some reach 60 feet (18 m), and it is
12–15 inches (30–38 cm) below the surface of the

ground. At the end, the female excavates a nesting chamber and lines it with wet grass and leaves. She carries this material by scooping it into a bundle that she wraps her tail around.

Usually two soft-shelled white eggs are laid, each ½ inch (1.3 cm) in diameter. The eggs often stick together, which prevents them from rolling away, and the wet leaves and grass keep them from drying out. Before retiring to lay her eggs, the female blocks the tunnel at intervals with earth up to 8 inches (20 cm) thick, which she tamps into position with her tail. During the incubation period of 10–12 days she rarely leaves the nest, but each time she does so these earth blocks are rebuilt. Presumably this is a defensive measure, although the platypus has virtually no natural predators except for occasional attacks by carpet snakes or goannas. This would suggest that in the past natural predators existed in some numbers and the platypus's earth block defenses were vital. The construction of earth blocks is an example of what is known as fossil behavior.

The young platypus is naked and blind, and its eyes do not open for 11 weeks. It is weaned at nearly 4 months old, when it takes to the water. The mother has no teats; milk oozes through slits on her abdomen, where it is licked up by the babies. A platypus matures at about 2½ years and has a life span of about 13 years in the wild.

Competing with rabbits

Formerly hunted ruthlessly for its beaverlike pelt, the platypus is now rigidly protected by law. However, it frequently becomes caught acci-dentally in wire cages set underwater for fish. A platypus that enters such a cage inevitably finds that it cannot escape again and drowns because it is not able to stay underwater for much longer than 5 minutes.

The rabbit, *Oryctolagus cuniculus*, which was introduced to Australia, threatens the platypus in a different way. In areas where rabbits have made many tunnels, the platypus cannot breed, as it needs undisturbed soil for its breeding burrows. However, although reduced in numbers, it is now well protected by the Australian authorities and is in no danger of extinction.

Creature of contrast

The fact that the platypus, a mammal, lays eggs and suckles its young was not known when the first specimen was discovered. In 1884, the British naturalist W. H. Caldwell, who had gone to Australia specially to study the platypus, dissected a female that had already laid one egg and was ready to lay another. Caldwell was thrilled by this discovery, and immediately telegrammed members of the British Association for the Advancement of Science, then meeting in Montreal, with the message "Monotremes oviparous, ovum meroblastic" (monotremes egg-laying, egg only partially divides). The term *monotremes* refers to the spiny anteaters and the platypus, both of which are egg-laying mammals and the only members of the order Monotremata. Delegates stood and cheered at this news, for controversy over this point had raged in the scientific world for some years.

The platypus's bill is equipped with nerve endings, enabling it to probe river and lake beds for food. Electroreceptors in the bill also help the platypus to detect prey.

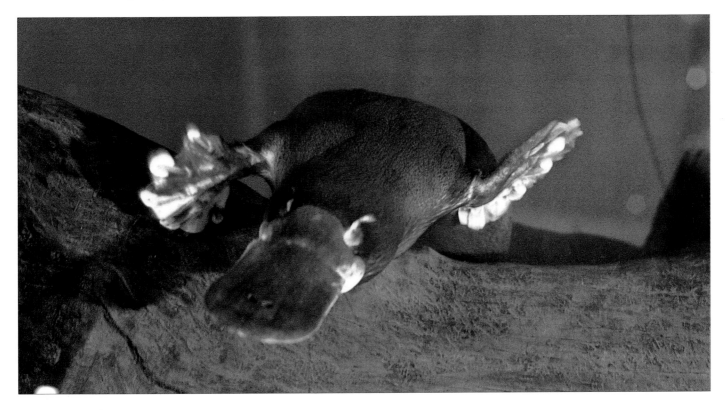

PLOVER

The ringed plover is one of the sand plovers, and is named for the black bands on its white forehead. It breeds on sandy or pebbly shores and often builds its nest just above the tide line.

THESE ATTRACTIVE SHOREBIRDS, which can often be seen running rapidly along the shore, are part of the large family of shorebirds that includes the lapwings, dotterels and killdeer. Here the plovers, genus *Pluvialis,* and the sand plovers, genus *Charadrius,* are discussed.

The three species of golden plovers in the genus *Pluvialis,* in addition to the black-bellied plover, *P. squatarola,* are distinguished by their spangled black and gold backs. Their underparts are usually black in the breeding season. For example, in breeding plumage the male Eurasian golden plover, *P. apricaria,* which is about 6¾ inches (17 cm) long, is spangled above and black below. It has a white line running from above the base of the bill, over the eye and following the wing margin. The female is similar. In winter the black disappears and the underparts become white with some mottling. The Eurasian golden plover lives in Iceland, the Faeroes, Britain, parts of northern Europe and northeastern Siberia.

The closely related American golden plover, *P. dominica,* is larger, up to 10½ inches (26 cm) long, but otherwise is similar to the Eurasian

golden plover. It breeds in the Arctic regions of Canada and Alaska. The black-bellied plover also breeds in the Arctic, but not in Greenland or Scandinavia. The breeding plumage of the male is silver, rather than gold spangled. The fourth species is the Pacific golden plover, *P. obscurus,* which looks similar to both the American and the Eurasian species, but is slightly smaller than both. It tends to be even more of a coastal bird than the other golden plovers.

The sand plovers are smaller, generally brown or gray above and white below, with black bands across the breast and face. They include the semipalmated plover, *Charadrius semipalmatus,* which breeds across large parts of Canada and Alaska, and spends the winter months around the coasts of Central and South America and the southern United States. There is also the ringed plover, *C. hiaticula,* of northern Europe, northern Asia, Greenland, Iceland, Spitzbergen and parts of Arctic Canada. The little ringed plover, *C. dubius,* is found over most of Europe, North Africa and Asia as far as Japan and New Guinea. The snowy or Kentish plover,

AMERICAN GOLDEN PLOVER

CLASS **Aves**

ORDER **Charadriiformes**

FAMILY **Charadriidae**

GENUS AND SPECIES ***Pluvialis dominica***

WEIGHT
Average 5 oz. (145 g)

LENGTH
**Head to tail: up to 10½ in. (26 cm);
wingspan: 25½–28⅓ in. (65–72 cm)**

DISTINCTIVE FEATURES
**Breeding: black belly, throat and face;
black upperparts with gold spangles; white
stripe above eye and on side of neck.
Nonbreeding: duller grayish upperparts;
gray underparts flecked with buff; white
stripe above eye.**

DIET
**Insects, arthropods, earthworms and other
invertebrates; berries in spring and autumn**

BREEDING
**Age at first breeding: 1 year; breeding
season: late May–July; number of eggs:
usually 2 to 5; incubation period: about 28
days; fledging period: about 30 days;
breeding interval: 1 year**

LIFE SPAN
Up to 30 years in captivity

HABITAT
**Breeding: well-drained Arctic tundra;
winter: inland grassland up to 3,940 ft.
(1,200 m) and coastal mudflats and river
margins during migration**

DISTRIBUTION
**Breeding: Arctic Canada and Alaska;
winter: central South America**

STATUS
Locally common

American golden plover ☐ breeding ☐ winter

*A black-bellied plover in winter plumage, Florida. Species of the genus **Pluvialis** are distinctive for their spangled black and gold backs and black underparts in the breeding season. Much of this dark coloration disappears in winter.*

C. alexandrinus, breeds in North America, the Caribbean, Europe, many parts of Africa, Asia and Australia. It no longer breeds in Kent or anywhere else in Britain. The most unusual of the plovers, in a genus all of its own, is the wrybill, *Anarhynchus frontalis,* of New Zealand. It is the only bird with a bill that is bent sideways.

Shore dwellers
Plovers spend most of their lives on shores or very open country, usually near water. However, the mountain plover, *C. montanus,* of the dry plains of the southwestern United States never lives near water and the golden plovers usually breed on well-drained tundra or high-altitude uplands. Outside the breeding season plovers gather in flocks to feed on estuaries, coastal mudflats and river margins, flying together with the precision often seen in flocks of shorebirds.

Insects, invertebrates and berries
Plovers eat mainly small animals such as insects, arthropods, worms and small mollusks. Sometimes one tries to tackle too much, for ringed plovers and others have been found with their

bills caught between the shells of cockles. The wrybill uses its unusual, sideways curved bill for probing under stones in search of insects. On migration the American golden plover eats large numbers of grasshoppers and locusts. Berries are also eaten by plovers, especially by the this last species. The American golden plover lives on the previous year's berries when it first arrives in the Arctic during spring, before the insects and other invertebrates have started to appear.

Careful guardians

Plovers' nests are usually no more than scrapes in the ground, without any lining. Many species regularly lay a clutch of four eggs but there may be as few as two, as in the wrybill, or as many as five. Both parents incubate and tend the young, which leave the nest shortly after hatching. The incubation period is about 4 weeks in larger species and 3 weeks in smaller species. The chicks fly after 3–5 weeks.

In the breeding season, plovers can be very inconspicuous, often being found only by searching for the source of their whistling calls. Their nests are also very difficult to find, and plovers regularly attempt to divert predators by distraction displays. In one such display the bird droops a wing as if it is injured, showing up the white plumage underneath. In another display the plover runs with its tail drooping. The ringed plover, moreover, spreads its tail to show a conspicuous white border as it runs away from a

A Eurasian golden plover incubating its eggs, Hardanger Vidda, Norway. In many plover species mating is monogamous, and both parents incubate and tend the young.

predator. These various displays distract the intruder, drawing attention away from the bird's well-camouflaged nest.

Long migrations

Many plovers make long migrations to and from breeding grounds. American golden plovers, for example, travel south from Arctic Canada and Alaska across Hudson Bay to Newfoundland and New England, and then across the western Atlantic to South America. A smaller number of southward migrants, chiefly juveniles, use the east and west coasts or an inland route across the Great Plains. In spring they return via a different route, through Central America and up the interior of North America. There is a reason for this circular tour: the plovers are taking the shortest route that allows adequate food on the way. In the fall they travel to the Atlantic coast to gorge on berries before setting off on their long sea journey to South America. The return trip would be impossible in spring because this area would still be foggy and frost-bound. The interior of the continent however, warms up quickly in spring and food will already be available.

Another interesting migration is made by the ringed plover. It has recently colonized Greenland, Ellesmere Island and Baffin Island. These populations do not migrate to South America but cross to Iceland and then follow their ancestors' trail south into Europe and West Africa.

PLUME MOTH

THE PLUME MOTHS make up a comparatively small family of moths in which the wings are divided into parts like feathers. The very long forewings are divided into two or three parts, and the hind wings are divided into three or four parts. A few plume moths have wings without clefts, such as the European genus *Agdistis*. The bodies of plume moths are very long and slender, as are the legs, which have prominent spurs. The wingspan may be ¼–2 inches (0.6–5 cm) and the flight is feeble. When plume moths are at rest, the forewings are rolled around the hind wings and usually are held straight out at right angles to the body.

Plume moths are found in most parts of the world. There are about 600 species, and those with common names are often named after their host plant (for example, the grape plume moth, *Pterophorus periscelidactylus*) or after their appearance (for example, the triangle plume moth, *Platyptilia gonodactyla*). Closely related are the feather-winged moths of the family Alucitidae, most of which are found in Asia. The many-plume moth, *Alucita hexadactyla*, is a member of the Alucitidae. It has a ½-inch (13-mm) wingspan and is found in Europe and the temperate parts of North America. The forewings and hind wings are divided into six and look like a pair of feather fans.

Cryptic creatures

Plume moths usually are active only in the evening or at night. During the day they make themselves very inconspicuous among the herbage or hedgerows, although a sudden glimpse may be caught of a disturbed plume moth as it flies out of a plant in which it has been resting. At rest, adults of *Agdistis* hold their rolled wings pointing forward in a line with the body, so they resemble the dried grass on which they sit. The moths' blackish markings, which look like molds, improve this camouflage.

Plume moths live on a variety of food plants but are only occasionally pests. *Exelastis atomosa* of India burrows into pea pods and eats the growing peas. The pupa is later formed within the pod. In the United States, the grape plume moth attacks the sprouting leaves of vines (genus *Vitis*), while the artichoke plume moth, *Platyptilia carduidactyla*, feeds on the globe artichoke, *Cynara scolymus*. As is usual in plume moths, the female lays the tiny eggs on the leaves of the artichoke plant, although she sometimes lays them on the flower buds. What the larvae feed on when they hatch depends on where the eggs were laid. Larvae hatching onto leaves some distance from flower buds mostly feed on the leaves, while those from eggs laid on leaves close to flower buds, or on the buds themselves, feed on the buds. The larvae start off pale yellow and molt three or four times into instars that increase in size with each successive molt. As they pass from one stage to another, the larvae feeding on the buds bore their way farther and farther into the buds, destroying them. The larvae do not pupate on the food plant but drop from it and find a suitable spot on the ground, often in leaf litter.

The leaves and sometimes the flowers of a plant are the usual food of plume moth caterpillars. Some, like those of the artichoke plume moth, bore into leaf and flower stems, making them wither; others bind the leaves into a rosette with silk strands so the caterpillars are protected while they eat. The caterpillars of *Stenoptilia pneumonanthes*, which feed on marsh gentian, *Gentiana pneumonanthe*, are sometimes drowned

The many-plume moth belongs to the family Alucitidae and is closely related to the plumed moths of the family Pterophoridae. This moth's larva is pink and feeds on the buds and flowers of honeysuckle (genus Lonicera*).*

GRAPE PLUME MOTH

PHYLUM	**Arthropoda**
CLASS	**Insecta**
ORDER	**Lepidoptera**
FAMILY	**Pterophoridae**
GENUS AND SPECIES	***Pterophorus periscelidactylus***

LENGTH
Adult wingspan: about ½ in. (1.4 cm); larva (caterpillar): up to ¾ in. (2 cm)

DISTINCTIVE FEATURES
Adult: narrow wings divided into featherlike plumes; very long, slender body; light brown overall, with lighter markings on wings; larva: light green with white hairs

DIET
Upper surface of grapevine leaves

BREEDING
Breeding season: late June–July; number of eggs: 2 to 10, laid in groups on vine; egg to adult: about 3–4 months, according to temperature and food quality; more than 1 generation per year possible

LIFE SPAN
Not known

HABITAT
Grapevines

DISTRIBUTION
Warmer regions of U.S.

STATUS
Common

The white plume moth, Pterophorus pentadactyla, *is found in Europe eastward to the Middle East. Its larva is green and yellow and feeds on bindweeds of the genera* Calystegia *and* Convolvulus.

by rain as they feed in the cup-shaped flowers. The caterpillars of *Agdistis bennetii* feed and hibernate on the leaves of sea lavender (genus *Limonium*), where they are in danger of being inundated by high tides.

There are often two generations of plume moths per year, the second hibernating as caterpillars in stems of the food plants. Plume moth pupae are soft and hairy and hang from leaves or stems like the chrysalises of some butterflies. Some species of plume moths weave a cocoon that is no more than a few strands of silk thrown around the body. The caterpillars of the triangle plume moth, the adult of which has black triangles on its forewings, hatch in June and feed on coltsfoot, *Tussilago farfara*. The adults emerge and are on the wing from August to September, then a second generation of caterpillars becomes active in September. They feed on the leaves of coltsfoot before burrowing into the stems to hibernate. In the spring they feed on buds and flowers and then pupate in the seed heads, binding the seeds with silk so they do not fall. The adults emerge in May.

Some plume moth pupae are capable of movement. The pupa of *Leioptilus lienigianus* jerks sharply if disturbed, an action that could perhaps startle a predator. That of *Platyptilia calodactyla* is also mobile. It often wriggles to the surface of its burrow in a leaf but retreats if disturbed. It is prevented from emerging too far by an anchor rope of silk, one end of which is attached to its hind parts, the other to the inside of the burrow.

Preying on the predator

Adults of *Trichoptilus paludum* can sometimes be seen during the day, flying over boggy heaths and moors. They lay their eggs on sundews (genus *Drosera*), bog plants that capture and devour insects. The upper surfaces of sundew leaves bear hairs surmounted by glands. When an insect lands on a leaf, it is trapped in sticky secretions. The hairs then bend toward the insect and secrete fluids that digest it. Somehow the larvae of *T. paludum* can feed on the sundew leaves without being caught and digested.

POCHARD

POCHARDS DIVE FOR their food, in the manner of grebes and loons, with a quick jump up before plunging under. Consequently, they are more adapted to an aquatic life than their dabbling relatives, the mallard, teal and pintail. Their legs are placed well back on the body, and most pochards have difficulty walking. They come ashore mainly for roosting and nesting. The short, heavy body has a large head and a short neck. The males have an eclipse plumage during midsummer, when they resemble the females.

Classification
The three pochards in the genus *Netta* are less adapted to an aquatic life and do not dive as well as the other species. They float higher in the water and walk more easily. The other 12 pochards are classed in the genus *Aythya*, of which there are four groups. The first of these includes the European pochard (*A. ferina*); the American canvasback (*A. valisineria*); and the redhead (*A. americana*). These are similar in plumage. The second group is made up of pochards called white-eyes, and includes the hardhead (*A. australis*) and the ferruginous duck or ferruginous pochard (*A. nyroca*). The third group of pochards are mostly black and white

and include the tufted duck (*A. fuligula*) and the New Zealand scaup (*A. novaeseelandiae*). The fourth group includes the scaups.

Lifestyle of the canvasback
The most important pochard is the canvasback, which breeds in western North America from Alaska to New Mexico. In the fall it migrates down the flyways as far as South America. The canvasback is found year-round in the northwestern United States, and winters on the Pacific coast as far south as Mexico and in most of the southern and central-eastern United States. Although it is not the most hunted duck, it is the species most esteemed as a food dish. Most of the male's body is light gray, with glossy black on the breast and russet on the head and neck. The female has a light brown head and breast and a pale eye stripe.

Male canvasbacks use a neck-stretching display to each other during competition for mates, a characteristic that they also employ to ward off intruders. Females also stretch their necks towards their chosen mate during courtship display, a gesture that is usually accompanied by a soft calling. Canvasbacks are monogamous, although the male deserts the female once the clutch has been completed. They breed on open

The red-crested pochard, Netta rufina *(male, above), is nearly the size of a mallard, but more brightly colored. It breeds mainly on reedy lakes and reservoirs in eastern Europe.*

The redhead favors shallow freshwater lakes, ponds and marshes, although it is also found on brackish coastal bays and lakes. Like other pochards, it dives to the bottom for its food.

freshwater marshes, ponds and lakes. The nest is built by the female out of aquatic plant materials, and floats on the water. Canvasbacks frequently desert their nests if flooding occurs before the female has had the opportunity to raise the nest above the water level with a pile of dead reeds and other plants.

Common predators of canvasbacks include raccoons, skunks, coyotes, foxes, mink, weasels, crows and magpies. Raccoons in particular cause much nest destruction during the canvasback egg-laying period. However, as well as natural predators, canvasbacks face the threat of habitat destruction from the draining of freshwater and prairie marshes for agricultural purposes. They are also at risk from lead poisoning, because they often eat lead shot in mistake for grain or seeds. In conjunction with overhunting, these factors have resulted in canvasbacks becoming the least abundant game duck in North America.

Related species

The European pochard is closely related to the canvasback. It is 16½–19⅓ inches (42–49 cm) long and breeds in Europe and Central Asia. The plumage of the two species is almost identical. However, the canvasback has a flatter forehead and a much longer, plain black bill, which creates a quite different silhouette.

The redhead is very similar in appearance to the canvasback, although it is smaller. The redhead sometimes lays its eggs in the nest of another duck, often that of a canvasback, as the two species have similar habitat preferences,

COMMON POCHARD

CLASS **Aves**

ORDER **Anseriformes**

FAMILY **Anatidae**

GENUS AND SPECIES *Aythya ferina*

ALTERNATIVE NAME
European pochard

WEIGHT
Usually 1¾–2¼ lb. (0.8–1 kg)

LENGTH
Head to tail: 16½–19⅓ in. (42–49 cm); wingspan: 28¾–32¾ in. (72–82 cm)

DISTINCTIVE FEATURES
Stocky body; short neck; high crown with long, sloping forehead; broad, gray bill. Male: chestnut brown head and neck; black breast and tail; rest of plumage pale gray. Female: yellow-brown overall.

DIET
Underwater vegetation; crustaceans, mollusks, aquatic insects, tadpoles and small fish

BREEDING
Age at first breeding: 1 year; breeding season: May–July; number of eggs: usually 8 to 10; incubation period: 24–28 days; fledging period: 50–55 days; breeding interval: 1 year

LIFE SPAN
Up to 10 years

HABITAT
Open fresh waters of medium depth, including reservoirs and ornamental ponds in Europe; occasionally along coasts

DISTRIBUTION
Summer: much of Europe, except north, and east across Central Asia and southern Siberia; winter: also in parts of Africa, India and southern Asia

STATUS
Common

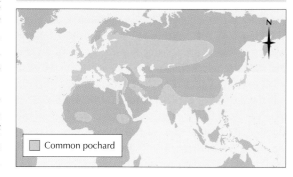

Common pochard

and the young are raised by foster parents. Such brood parasitism often results in the foster mother abandoning her nest. The redhead is distinguishable from the European pochard by its yellow iris and more rounded head. The male redhead's body is gray, apart from the black breast, while the female has a dull gray plumage.

The tufted duck, 15¾–18½ inches (40–47 cm) long, is an Old World bird, ranging from Iceland to Kamchatka. The male is all black with very distinctive white flanks and a small trailing crest. The greater scaup, *A. marila*, with a circumpolar Arctic distribution, has a plumage similar to that of the tufted duck, but lacks the tuft and has a light gray back. In North America this species is also known as the bluebill or broadbill.

Most pochards live in the Northern Hemisphere. Exceptions to this include the southern pochard (*Netta erythrophthalma*), which ranges from Ethiopia to South Africa, the New Zealand scaup and the hardhead.

Commuting to feed

Pochards migrate in V-formation; sometimes the whole flock plunges from a great height toward a sheet of water. Their habits are generally similar to those of dabbling ducks. Pochards breed in fresh water, although some spend the winter in salt water or the brackish water of estuaries, where they can be found in large flocks. Others go to sea only during very cold weather. They feed mainly in the morning and evening and sometimes at night. Shallower waters are preferred for feeding, and in the evening there is a retreat from larger and safer waters to smaller pools. Pochards are awkward when they take off or land, as they have to run over the water before gaining sufficient lift with their short wings.

Pochards feed on aquatic plants and animals, usually diving to collect them from the bottom. The canvasback eats mainly plant food. About three-quarters of the canvasback's food is vegetable, including water lilies and wild oats, although it also takes worms, mollusks, tadpoles, crustaceans and a few fish. Canvasbacks particularly favor the smallage, or wild celery, that grows around Chesapeake Bay. This has long, ribbonlike leaves and grows just offshore in great quantities. The canvasbacks prefer the roots, which taste like garden celery. They dive to uproot the plants and bite off the roots, leaving the remainder to float away. Redhead and scaup also dive for water celery roots. Some pochards swim underwater with their wings, staying under for up to 1 minute or more. Both European pochards and tufted ducks prefer to feed in shallow water 2–6 feet (60–180 cm) deep, but scaup also feed in deep water.

Nesting and rearing the young

Pochards nest on marshes or lakes and ponds where there is abundant cover around the edges. Nests are often quite close together and usually are very near water. Canvasbacks and European pochards may nest among reeds in the water.

The female broods the 6 to 12 eggs, rarely leaving the nest during the incubation period, which may last up to about 28 days. The male usually leaves the female before the chicks are hatched, although the male greater scaup sometimes stays with his family. The chicks are led from the nest as soon as they are dry and are guarded and fed by the female, which brings food to the surface for them before they have learned to dive. The canvasback abandons her brood early, and these broods band together, forming groups of first-year ducks.

The plumage of the male common pochard lasts only for the breeding season. Thereafter the bird takes on the dull brown coloration of the female.

POCKET GOPHER

The pocket gopher's sturdy, tapering body, long incisors, powerful forelimbs and robust claws make it a highly efficient burrower. Pictured is Botta's pocket gopher, Thomomys bottae, which ranges from California and Nevada south to Mexico.

THE NAME OF THESE small hamsterlike animals is derived from the two external fur-lined cheek pouches, which run from face to shoulder. They are used for carrying food and are turned partially inside out when they need cleaning. There are 40 species of pocket gophers, which vary considerably in size, with a body length of 4½–14 inches (11.5–35 cm) and a tail 1½–5½ inches (4–14 cm) long. The coat is gray brown or black and the fur color varies from almost black through many shades of brown to off-white; albinos are common. Pocket gophers living in hot lowland areas have shorter, coarser fur than those in cooler climates. Those species belonging to the genera *Macrogeomys* and *Zygogeomys* have characteristic white patches.

Pocket gophers are adapted for burrowing. The skull is large and angular, and the body is thickset and tapers toward the tail. Ears and eyes are small, the latter being kept moist with a thick fluid to keep the eyeball free from dirt. The legs are short and powerful, specially the forelegs, which have long digging claws. The tail is almost naked and is very sensitive to touch. By arching the tail to raise the tip just off the ground, a pocket gopher can feel its way as it moves rapidly backward in its burrow. All pocket gophers have long, curving upper incisors. The lips close behind them to stop earth entering the mouth while they are burrowing.

Pocket gophers are native to North America and range from western Canada south to Panama. They do not travel far and are found in localized areas, often restricted to valleys by mountain barriers.

Solitary burrowers

Pocket gophers spend almost their entire life underground, only very occasionally coming to the surface to collect food. Each gopher lives within its own system of burrows and they only come together at mating times. Young pocket gophers sometimes venture aboveground after leaving their parents. Adults may be driven out of their burrows by drought or after floods and be forced to search for new homes.

The individual burrows are often extensive and are marked by fan-shaped mounds of earth around the entrances. These are carefully blocked with earth, both as protection against predators and to maintain suitable temperature and humidity. There are two types of tunnels: long, shallow ones, used mainly for getting food, and deep ones, used for shelter, with separate chambers for food storage, nesting and latrines. A large assortment of other animals use inhabited and abandoned gopher burrows.

Pocket gophers dig with their strong foreclaws, using the curved incisors to loosen hard earth and rocks. In common with other rodents, gophers' incisors grow continually, and worn surfaces are regularly replaced. The incisors may grow as much as 20 inches (50 cm) in one year.

Pocket gophers do not hibernate but may remain relatively inactive for long periods. Evidence of their winter activity is apparent when the snows melt in high pastures to reveal gopher burrows within the snow, lined with earth and left behind as a mass of crisscrossing cores showing that the animals have burrowed at several levels. The tunnels are up to 40 feet (12 m) long and 2–3 inches (5–7.5 cm) across.

Underground vegetarians

The staple diet of pocket gophers is tubers, bulbs and roots of plants that can be found and eaten in the security of the burrow. Occasionally at night or on overcast days, pocket gophers surface in search of food, but more often the rodents feed from the safety of their burrows, pulling down plants by cutting the roots and

POCKET GOPHERS

CLASS	**Mammalia**
ORDER	**Rodentia**
FAMILY	**Geomyidae**
GENUS	***Geomys, Thomomys, Zygogeomys Pappogeomys, Cratogeomys, Orthogeomys***
SPECIES	**40 species**

ALTERNATIVE NAMES
Gopher; tuza

WEIGHT
Most species 6–18 oz. (150–450 g)

LENGTH
Head and body: 4½–14 in. (11.5–35 cm); tail: 1½–5½ in. (4–14 cm)

DISTINCTIVE FEATURES
Sturdy, hamsterlike body; strong forelimbs and claws; 2 large cheek pouches; few hairs on tail; gray-brown or black fur; light-colored face patches and white marks on body in some species

DIET
Mainly roots, tubers, bulbs and stems

BREEDING
Age at first breeding: 2–4 months; breeding season: all year; number of young: 1 to 8; gestation period: 18–52 days; breeding interval: 2 to 4 litters per year

LIFE SPAN
Up to 7 years

HABITAT
Deserts; lightly wooded forests and prairies; agricultural land

DISTRIBUTION
Southern Canada south to Panama

STATUS
Most species common; critically endangered: *O. cuniculus*; endangered: *T. umbrinus*; vulnerable: *P. alcorni, G. personatus*; at low risk: *C. tylorhinus, C. zinseri*

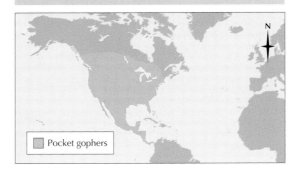

Pocket gophers

pulling on the stem. In agricultural regions, crops often suffer badly from the gophers' actions: the rodents favor sweet potatoes, sugarcane, peas and fruits. Gophers rarely drink. It is likely that they obtain sufficient moisture from plant food.

With suitable soil and food conditions, pocket gopher numbers increase rapidly. A female may have two to four litters a year, usually in the spring and late summer, each numbering one to eight offspring. The newborn young, weighing about 2.4 grams, are blind and almost hairless. In most species they leave their mother at about 2 months to make their own burrow network and are sexually mature within 4 months.

Subterranean safety

The best protection a pocket gopher has against its predators is its underground lifestyle, but in fights with its own kind it is protected by the loose skin and thick hair around the head. Predators, such as coyotes, badgers and skunks, sometimes succeed in digging pocket gophers out, and if the rodents come aboveground for any length of time, owls and hawks take them. Snakes and weasels also hunt pocket gophers in their burrows. Humans are the major persecutor of pocket gophers; the rodents can be very destructive, eating crops, burrowing through dykes and contributing to soil erosion. However, they do improve the soil by loosening, aerating and mixing it with organic matter. They also conserve water: after a heavy snowfall, the melted water sinks deep into the earth through the maze of gopher tunnels instead of flowing straight into the nearest stream.

A northern pocket gopher, Thomomys talpoides. *Pocket gophers usually leave their burrows only to find a mate or to dig a new burrow if their old one is damaged. When above ground they are extremely vulnerable to predators.*

POLAR BEAR

Their broad, hairy feet provide polar bears with a good grip on ice. When they are pursued the bears can run at speeds of 18–24 miles per hour (30–40 km/h).

THE POLAR BEAR IS THE largest land carnivore. The males average 6½–8¼ feet (2–2.5 m) long and may reach 9 feet (2.7 m) standing erect on their hind legs, with a shoulder height of about 5¼ feet (1.6 m). Male polar bears may weigh up to 1,720 pounds (780 kg), although the females are smaller, with an average weight of about 660 pounds (300 kg). Polar bears have a long head with small ears and a straight nose, a long neck, powerful limbs, broad feet with hairy soles and a stumpy tail 3–5 inches (7.5–12.5 cm) in length. Although it appears to be white, polar bear fur consists of hollow, colorless hairs. These conduct heat to the polar bear's skin, which is black and therefore a highly efficient heat absorber. The coat becomes more yellow in the summer, and with age.

Polar bears live along the southern edge of the Arctic pack ice. They are carried southward by the ice in spring and summer and return northward when the ice breaks up.

A home of ice and water

Polar bears are expert divers and swim strongly, reaching a maximum speed of about 7 miles per hour (12 km/h), using only the front legs and trailing the hind legs. A thick layer of fat under the skin, 3 inches (7.5 cm) thick on the haunches, helps to insulate the bears and to keep them buoyant in the water. Polar bears have been seen swimming strongly 200 miles (320 km) from land and can cover more than 62 miles (100 km) in one go. The bears usually swim with the head stretched forward, but when the sea is rough they put their heads underwater, lifting them periodically to breathe. When polar bears come onto land, they shake themselves dry as dogs do. They swing their heads from side to side as they walk, as if searching or smelling out prey. The bears are essentially nomadic, and wander for miles in search of food.

Hunting seals

Polar bears' favorite food is seals, especially the ringed seal, *Pusa hispida*, which the bears stalk by taking advantage of snow hummocks. A seal asleep on the edge of the ice is easy prey. The polar bear swims underwater to the spot, comes up beneath the seal and crushes its skull with one blow of the powerful forepaw. When a ringed seal is about to give birth, she digs an igloo in a hummock of snow over her breathing hole. Polar bears sniff out seal igloos and take the pups and also catch seals when they come up for air at breathing holes. Polar bears kill young walrus, *Odobenus rosmarus*, but in a fight with a grown walrus the bear is likely to lose. The bears also eat fish, seabirds and their eggs and carrion.

POLAR BEAR

CLASS **Mammalia**

ORDER **Carnivora**

FAMILY **Ursidae**

GENUS AND SPECIES **_Ursus maritimus_**

ALTERNATIVE NAME
Ice bear

WEIGHT
Up to 1,720 lb. (780 kg), male much larger than female

LENGTH
Head and body: 6½–8¼ ft. (2–2.5 m); shoulder height: about 5¼ ft. (1.6 m); tail: 3–5 in. (7.5–12.5 cm)

DISTINCTIVE FEATURES
Large, powerful bear; long head and neck; small ears; broad feet with hairy soles; stumpy tail; thick white coat, becoming cream colored with age; dark nose

DIET
Mainly seals and carrion; occasionally young walrus and musk-ox; also fish, lemmings, vegetation and berries

BREEDING
Age at first breeding: 10–11 years (male), 5 years (female); breeding season: March–April; number of young: usually 2; gestation period: 195–265 days, including delayed implantation; breeding interval: usually 2–4 years

LIFE SPAN
Up to 30 years

HABITAT
Summer: Arctic tundra grassland; winter: frozen pack ice

DISTRIBUTION
Arctic coasts

STATUS
Conservation dependent

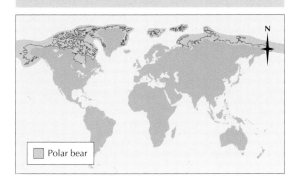

Polar bear

A stranded whale draws bears from a large area. At certain times of the year, usually in late spring or early summer, polar bears eat large quantities of grass, lichens, seaweed and moss. They also take crowberries, bilberries and cranberries.

The cubs of polar bears are able to walk around 47 days after their birth, but are not fully weaned until they are 3 months old.

A long gestation

Mating is in April or May. The implantation of the fertilized egg in the uterus, and its development, are delayed. Consequently, polar bears have a fairly long gestation period of 195–265 days. Two cubs are usually born in December or January. At birth each cub is 1 foot (30 cm) long and weighs 1½ pounds (0.7 kg). It has a coat of short, sparse hair. The eyes open at 33 days, and the ears open at 26 days, although hearing is imperfect until the cub is 69 days old. Male polar bears are sexually mature at about 10 years, although females mature in about half this time.

It is often said that cubs are born while the mother is hibernating. Polar bears, like other bears, do not hibernate in the strict sense, and it is now usual to speak of their sleep as winter dormancy. The pregnant she-bear seeks out a bank of snow in the lee of a hill and digs into it. There is some scientific debate as to whether the males hibernate. It seems likely that some do, although for shorter periods than the females.

Humans are the main threat

Apart from humans, the polar bear has no predators, although a walrus may gore a polar bear in self-defense. Young bears die of accidents, by drowning or by being crushed by ice during storms; old males also kill and eat cubs. Polar bears have long been hunted by Inuits for their meat and pelts. Their long canine teeth are also used for ornaments. Indigenous peoples also make bed covers, sleigh robes and trousers from polar bear pelts.

Although Inuits eat the flesh of polar bears, they do not make use of the liver, even to feed their dogs. It is poisonous, due to the fact that it contains high concentrations of vitamin A, which causes headaches and nausea and sometimes a form of dermatitis (skin inflammation).

Conserving the polar bear

The earliest known record of polar bears being taken into captivity may date from 880 C.E., when two cubs were taken from Iceland to Norway. At the time polar bears were often offered as gifts to European rulers, who rewarded the donation with valuable goods or titles.

Intensive hunting of the bears began in the 17th century, when whalers reached the Arctic pack ice. Two centuries later the population of polar bears was decreasing in many parts of the Arctic. Subsequently, sealers also made an impact on polar bear numbers, and in 1942 alone Norwegian sealers killed 714 bears. More recently the Inuits have hunted polar bears in

order to trade their pelts. In parts of arctic North America it was relatively easy for hunters to land by airplane and shoot the bears for sport in the past, but legislation controlling this is now in place and under fairly constant review.

In 1965, the world population of polar bears was estimated at over 10,000, with a total annual kill of about 1,300. In 1968 research scientists from five Arctic nations—the Soviet Union, the United States, Canada, Norway and Denmark—established the Polar Bear Group to study the life history, ecology, seasonal movements and population of the species. As a result of the work of this group, the International Agreement on the Conservation of Polar Bears and their Habitat was signed between the five participating nations in 1973. The agreement prevented the hunting and capture of polar bears unless for strictly defined purposes and banned the use of motorized vehicles or aircraft to hunt them, although indigenous peoples were permitted hunting rights for subsistence purposes. It also outlawed the sale of the skins of any polar bears caught within the agreed conditions and stated that any ecosystems polar bears belong to should also be protected. Polar bears are common in areas where they aggregate, but are generally rare. Contemporary estimates put the current polar bear population at 21,000 to 28,000.

Polar bear cubs stay with their mother until they are at least 10 months old. When she is ready to mate again, the mother drives away any cubs that still remain with her.

POLECAT

POLECATS AND FERRETS are frequently regarded as the same species. It is likely that the European polecat, *Mustela putorius,* is the ancestor of the ferret, a fact reflected in the ferret's Latin name, *M. p. furo,* although theoretically the ferret may be a descendant of the steppe polecat, *M. eversmanni.* All three belong to the same genus as the North American black-footed ferret, *M. nigripes,* discussed elsewhere in this encyclopedia. Scientists believe that the ferret was domesticated from wild polecats by humans about 2,500 years ago, to hunt small animals, mainly rabbits, although it is uncertain which ancient people first used them for this purpose. The marbled polecat, *Vormela pergusna,* is sometimes regarded as a subspecies of *M. putorius.*

The genome (genetic structure) of polecats and ferrets is very similar, although there are differences in bodily structure. The shape of the skull differs between the two, and the teeth of the ferret are more closely packed and more variable in number than those of the polecat. In addition, there are differences in eye structure, leaping and balancing abilities and the method each animal uses to locate sounds.

The male European polecat grows up to 25½ inches (64 cm) long, of which 4¼–7½ inches (10.5–19 cm) is bushy tail, and weighs up to 3¾ pounds (1.7 kg). The female is slightly smaller and only a little more than half the weight of the male. In both sexes the long, coarse fur is dark brown on the upperparts, with dense yellowish underfur showing through, and black on the underparts. The short legs and tail are black, as is the head, which has white or yellow patches between the small ears and eyes from June to November.

The polecat is fairly common or uncommon throughout Europe as far north as southern Sweden and southern Finland. In Britain, however, it has become very rare except in Wales. Indeed, in mid-Wales its numbers seem to be increasing. The unpleasant discharge from its glands gave the polecat, sometimes called the fitchew, its former English name of foumart or foul-marten. Its fur, sold under the name "fitch," was widely used in the early 19th century.

Streamlined and stealthy

The polecat's long, almost cylindrical body enables it to move at speed as it slinks across bare ground or close-cropped grass. It generally moves with its body slung so low that it almost seems to be touching the ground. The long neck is stretched out in motion and the short legs move almost in a paddling action, so the animal

The polecat's long, dense fur helps it to withstand the low temperatures of the northern extremes of its range. Fur traders hunted the animal for its coat until about 100 years ago.

Play helps young polecats to develop hunting skills they later use as adults.

EUROPEAN POLECAT

CLASS	**Mammalia**
ORDER	**Carnivora**
FAMILY	**Mustelidae**
GENUS AND SPECIES	***Mustela putorius***

ALTERNATIVE NAMES
Fitchew, foumart, foul-marten (archaic, U.K. only)

WEIGHT
16–68 oz. (0.4–1.7 kg), male much larger than female

LENGTH
Head and body: 12–18 in. (30–45 cm); tail: 4¼–7½ in. (10.5–19 cm)

DISTINCTIVE FEATURES
Streamlined, lithe body; short legs; bushy tail; dark brown pelage, with paler underfur; head, feet and tail darker than rest of body; yellow spots around temples; pale muzzle

DIET
Mammals up to the size of rabbits; also frogs, lizards, snakes, invertebrates, bird eggs, vegetation and berries

BREEDING
Age at first breeding: about 1 year; breeding season: March–May; number of young: 4 to 6; gestation period: about 42 days; breeding interval: 1 year

LIFE SPAN
Up to 6 years

HABITAT
River edges, woodland and farmland

DISTRIBUTION
Western Europe east to western Russia and Caucasus region

STATUS
Fairly common or uncommon; very rare in Britain, but population increasing in Wales

European polecat

seems to glide rather than run. Seen from the side, the action of the limbs is more reminiscent of swimming than of running.

If they are domesticated at a young age, polecats are easily tamed and give little sign of their traditional smell, probably because this is used in the wild to mark a territory, not, as was formerly thought, as a defense. Polecats also deposit urine and feces along their trails to mark them. Young polecats spend a lot of time playing together, during which they frequently grip each other by the back of the neck. It is likely that such play enables them to develop the neck bite used to kill their prey.

Versatile predator

The polecat flourishes in a great variety of habitats and is as much at home among sand dunes and on sea cliffs as it is on lowland farms or in wooded gorges. However, it prefers woods and copses, making its den in suitable holes, such as a fox earth, rabbit burrow or natural rock crevice.

With the approach of winter, polecats may seek shelter in deserted buildings. They are less agile than martens, and cannot climb as well as martens do. Polecats are active mainly at night, depending on their keen sense of smell to hunt down prey. They are usually silent but occasionally use short yelps, clucks and chatterings. They also scream and hiss when frightened, as part of a defensive threat and to communicate.

Polecats feed on rats, mice and rabbits, birds and their eggs, frogs, lizards and snakes. Because they are smaller than the males, female polecats are able to enter the burrows of small prey, such as rats and voles, that males cannot enter. Male polecats tend to hunt larger prey, such as rabbits.

For this reason, although polecats are territorial animals, males seek to exclude only other males from their territories; as females do not hunt the same prey as males, they are not a threat and so are tolerated. Similarly, female polecats attempt to exclude other females from their range, but tolerate the presence of males. In contrast to the solitary and territorial polecats, ferrets are gregarious animals.

Polecats are reputed to be major predators of farmyard poultry. Indeed, the name polecat is said to come from the French *poule chat*, meaning "chicken cat." Consequently, polecats and other mustelids have been widely hunted as pests, and the drastic reduction in numbers of the European polecat in England was probably due to intensive hunting by gamekeepers. However, the polecat may also help to keep the populations of small mammals, especially rats and mice, within bounds. It is possible that the wholesale reduction in the numbers of polecats might have contributed in the past to the increase in the populations of rats.

Small mustelids have often been regarded as a useful form of pest control by farmers. The European common weasel (*M. nivalis*), the ermine (*M. ermina*) and the ferret were introduced to New Zealand in 1884 in an attempt to control the numbers of rabbits that were then over-running the country's sheep pastures. However, they did not succeed in keeping rodent numbers down. Prey such as rabbits reproduce at a far more rapid rate than their predators, such as polecats. Predation is usually most successful when prey numbers are in decline, a situation that may be brought about by a shorter breeding season or by an increase in the age of sexual maturity, for example. In conditions such as these, predation by small mustelids may accelerate or extend a decrease in the population of rodents.

White young

The polecat mates in early spring, probably in April. Following a gestation period of about 6 weeks, litters of up to 11 young, generally 4 to 6, are born in a nest of dry grass in woods or among rocks. The young are blind for their first fortnight and are pure white, the colors and markings of the adults appearing only after about 3 months.

Some naturalists believe that there is probably a second litter a few months later. However, this is unlikely because the young do not leave their parents until they are 3 months old, pushing the time of a second litter to late summer or early fall. Young polecats huddle in the nest to keep warm, but in hot weather they lie as far from each other as possible.

The European polecat is found in a wide variety of habitats, from sand dunes to hills. However, it favors woodland, farmland and river edges.

POLLACK

Pollack are coastal members of the cod family and these were photographed off Cornwall, southwest England. The origins of the name pollack *are not known, and it has been variously altered in the past to pollock, podlok and podley.*

A PURELY EUROPEAN MEMBER of the cod and haddock family (Gadidae), the pollack, *Pollachius pollachius*, is the closest relative of the coalfish, *Pollachius virens*. The latter is found on both sides of the Atlantic and is known as the pollack in North America. For the purposes of this article, however, the term *pollack* refers only to *Pollachius pollachius*.

The pollack is probably the most handsomely colored member of the Gadidae. It is dark brownish green on the back, shading to light green on the sides, which are streaked and spotted with yellow. The belly is white. The fins are dark yellowish green and the eye is golden. The lateral line is dark and strongly curved over the pectoral fin.

One way of recognizing the pollack is by the jutting lower jaw. Also, unlike some other members of the Gadidae, such as the Atlantic cod

(*Gadus morhua*) and the haddock (*Melanogrammus aeglefinus*), the pollack has no barbel on the chin. Otherwise its shape conforms to the typical cod pattern, with 3 dorsal and 2 anal fins, small pectoral fins and very small pelvic fins lying under the throat. Pollack are up to 4¼ feet (1.3 m) long and up to 26½ pounds (12 kg) in weight. They are found in the northeast Atlantic from Scandinavia south to the Bay of Biscay.

Midwater feeders

Pollack are more a coastal fish than any other member of the cod family. They are often taken by line fishers from the shore, and when hooked they dive powerfully for refuge among the rocks. The pollack is not highly esteemed for its flesh, although it is regarded as a good sport fish. It is plentiful, especially off Scotland, where it is known as the lythe. Pollack come to the surface

POLLACK

CLASS	**Osteichthyes**
ORDER	**Gadiformes**
FAMILY	**Gadidae**
GENUS AND SPECIES	***Pollachius pollachius***

ALTERNATIVE NAMES
Lythe (Scotland only); pollock, podlok, podley (archaic)

WEIGHT
Up to 26½ lb. (12 kg)

LENGTH
Up to 4¼ ft. (1.3 m)

DISTINCTIVE FEATURES
Body form typical of cod family, with 3 dorsal and 2 anal fins but no chin barbel; lower jaw protrudes beyond upper jaw; dark lateral line, strongly curved over pectoral fin; dark brownish green on back, shading to yellowish green on sides; white belly

DIET
Adult: mainly fish; also squid, cuttlefish, shrimps and crabs. Young: bottom-living worms, crustaceans and mollusks.

BREEDING
Breeding season: January–April; number of eggs: up to 2.5 million

LIFE SPAN
Up to 8 years

HABITAT
Mostly close to shore over firm, rocky seabeds; moves to deeper water in winter

DISTRIBUTION
Northeast Atlantic: Norway, Faeroe Islands and Iceland south to Bay of Biscay

STATUS
Locally common

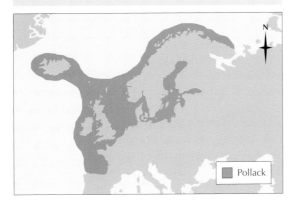

Pollack

at night and are readily attracted by lights. Mackerel fishers often catch them with bright, moving baits. Their preference for moving bait gives a clue to their food. They feed in midwater on small fish, especially sand eels (family Ammodytidae), sprats, herrings and pilchards (all of the family Clupeidae). Pollack also consume squid, cuttlefish, shrimps and crabs as and when they come across them.

Junior marathons

Pollack spawn from January to April. Their spawning grounds are in deep water and the females lay up to 2.5 million each, at depths of 330–660 feet (100–200 m). The eggs measure just over 1 millimeter in diameter and float near the surface. Both they and the larvae are pelagic but gradually drift shoreward, and young pollack in their first year are particularly common close inshore. The spawning grounds are small and widely scattered: as far apart as the western English Channel and the Hebrides islands off the west coast of Scotland. Because of this, young pollack make long migrations as they spread out around the coasts. They feed on the seabed on crustaceans, worms and mollusks.

The pollack exhibits the typical body form of the family Gadidae. This commercially important group of fish also includes the coalfish, the haddock and the Atlantic cod.

POMPADOUR FISH

Popular aquarium fish because of their bright colors and attractive patterns, pompadour fish are actually quite difficult to keep and require frequent water changes. Pictured are a group of red discus, one of the two species of pompadour fish.

THESE FISH FROM THE RIVERS of the Amazon Basin have been described as the noblest among aquarium fish. They are known as the king of freshwater fish and by their popular name, pompadour fish, as well as by the descriptive name discus fish because of their shape. Pompadour fish are almost disc-shaped when fully grown and up to 8 inches (20 cm) long. The long dorsal and anal fins make the otherwise oval body look more nearly circular. The body is covered with small scales, but the cheeks and gill covers are more markedly scaly. The mouth is small, with thick lips. There is a single row of small, conical teeth in the middle of each jaw and instead of the usual two pairs, there is a single pair of nostrils.

The colors are not easy to describe because they change with age, and many color variations have appeared with selective breeding in aquaria. All young pompadour fish are brown, with several vertical, dark bars down each side.

At 6 months old, flecks of blue appear on the head and gill covers, and these spread until the sides are colored with alternating, horizontal bands of blue and reddish brown. There are nine vertical, dark bands, the first running through the eye. The fins become blue at their bases, pale blue and orange on the outer edges, and there are streaks of blue and orange between. The pelvic fins are red with orange tips.

There are two distinct species, the red discus, *Symphysodon discus*, and the blue discus, *S. aequifasciatus*. Color variations include the brown discus, the Peruvian green discus, the red dragon and various subspecies named from their breeders, including the Jack Wattley turquoise.

Prefer heavy vegetation

The red discus occurs in the Amazon Basin in Brazil near the mouth of the Rio Negro, in the lower Rio Abacaxis and in the lower Rio Trombetas. The blue discus is found in the

RED DISCUS

CLASS	**Osteichthyes**
ORDER	**Perciformes**
FAMILY	**Cichlidae**
GENUS AND SPECIES	***Symphysodon discus***

ALTERNATIVE NAMES
Pompadour fish; Heckel discus

LENGTH
Up to 8 in. (20 cm)

DISTINCTIVE FEATURES
Disc-shaped, strongly compressed body; small mouth with thick lips; single row of small, conical teeth. Adult: reddish overall with light blue fins; 9 vertical bars, 3 of which distinct dark blue; variable number of pale blue flecks and horizontal blue bands. Young: brown, with darker vertical bars.

DIET
Worms, crustaceans, insects and plants

BREEDING
Number of eggs: several hundred; hatching period: about 50 hours; adults care for both eggs and larvae

LIFE SPAN
Not known

HABITAT
Heavily vegetated backwaters and pools

DISTRIBUTION
Amazon Basin, Brazil

STATUS
Common

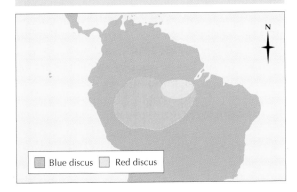

Blue discus ▢ Red discus

Amazon and Slimoes Rivers of Brazil, from the lower Rio Putumayo-Ica. It has also been introduced to the Rio Nanay in Peru. They are normally found in schools, but are territorial during the breeding season. The preferred habitats of these fish are backwaters and pools where there

is a lot of vegetation. They spend the day sheltering in the shadows of water plants when they are not feeding, and they avoid strong sunlight. Pompadour fish are omnivorous and feed on water insects, especially the larvae of midges and small dragonflies, small worms, similar invertebrates and some plant material.

Nursing the young

There is a brief courtship, during which the male and female pompadour fish clean the surface of a broad leaf of a water plant. When this is ready, the female lays rows of eggs, usually several hundred, on the leaf. Sometimes the surface of a stone is used, but only after being cleaned. Once the eggs are laid, the male swims over and fertilizes them. The parents take turns fanning the eggs with their fins, and the eggs hatch after about 50 hours.

Pompadour fish have an unusual form of parental care. As each baby breaks out of the egg, it is removed in the parents' mouth and placed on a leaf, where it hangs by a short thread for the next 60 hours or so. The parents continue to fan the fry (young fish) with their fins. Then the babies swarm at the side of one or other of their parents and attach themselves to its body or fins. The adults secrete a mucus-like substance onto the skin, on which the young feed for several weeks.

When 5–6 weeks old, the fry become independent and feed on small animal plankton such as very small water fleas or their larvae. At first they are the normal fish shape, if a little plump in the body. The discoid shape comes with age.

Pompadour fish, such as this blue discus, are known for their unusual parental care. For the first few weeks of life the young attach themselves to one of their parents and feed off a special slime secreted through the adult's skin.

POND SKATER

The pond skater is an efficient freshwater predator. Organs on the insect's legs enable it not only to detect the ripples sent out by potential prey in the water but also to work out in which direction the struggling animal is located.

POND SKATERS, OR WATER STRIDERS, are the familiar insects that can be seen floating or skimming swiftly over the surfaces of ponds, streams and flooded ditches. Their bodies are flat, narrow and ⅕–¾ inches (5–20 mm) long. They appear to have only two pairs of legs. The front pair are short and held close to the head, just behind the antennae. The rear two pairs are long and the tips are fringed with hairs, which rest on the surface of the water. Some species are wingless, others have small wings and a few have fully developed wings. Pond skaters develop through a series of molts from small nymphs. Adult bugs can usually be distinguished from nymphs by their fully developed wings, but adult pond skaters are not so easily recognized, because an individual with small wings could be a nymph or an adult, depending on the species.

Most pond skaters are freshwater insects, although one European species lives in estuaries and is also found on the coasts of the Baltic Sea. Pond skaters of the genus *Halobates* are among the very few marine insects. They live in tropical

and subtropical seas, sometimes hundreds of miles from land. True pond skaters belong to the family Gerridae, but other bugs that float on the surface of the water are sometimes called pond skaters. These include the water measurers, or water gnats, of the family Hydrometridae, which have a long, narrow head, and the water crickets of the family Veliidae. There are also the pondweed bugs of the family Mesoveliidae. One member of this family is terrestrial, living among fallen leaves in the forests of New Guinea.

Running on water

Pond skaters often gather in groups on the edges of ponds and lakes, scuttling over the water when disturbed and then regrouping. They propel themselves with the long pair of middle legs and steer with the hind legs. When pond skaters are moving slowly, each flick of the legs sends them sliding several inches over the surface of the water. However, pond skaters also hop across the water, rising 1 inch (2.5 cm) or so at each bound. Water plants present no obstacle, and pond skaters walk over floating weeds and

POND SKATERS

PHYLUM	**Arthropoda**
CLASS	**Insecta**
ORDER	**Hemiptera**
FAMILY	**Gerridae**
GENUS	***Gerris, Halobates*; others**
SPECIES	**About 350, including *Gerris lacustris*, *G. remigis* and *G. thoracicus***

ALTERNATIVE NAME
Water strider

LENGTH
⅕–¾ in. (5–20 mm)

DISTINCTIVE FEATURES
Flattened, narrow body; front pair of legs is short and held close to head; middle and rear pairs of legs are long, strong and fringed with hairs at tips; some species winged, others wingless

DIET
Small insects falling on water

BREEDING
Hemimetabolous (pass through incomplete metamorphosis). Breeding season: spring; number of eggs: about 100; 2 generations per year.

LIFE SPAN
Usually few weeks to months

HABITAT
Surface of still, freshwater ponds, streams, canals and ditches; *Halobates* species on floating seaweed and on seawater

DISTRIBUTION
Almost worldwide

STATUS
Most species common

even over dry water-lily leaves. Pond skaters feed on dead or dying insects that fall onto the water. They grasp the prey with their front legs and pierce the body with their mouthparts, sucking it dry. Cannibalism is common. Prey is detected by sight or by sense organs on the legs that pick up vibrations.

Adult hibernation

In temperate regions there are two generations of pond skaters each year. The second generation hibernates on land, occasionally taking to the water on warm winter days. In the spring and early summer the hibernating pond skaters return to water and mate. About 100 eggs are laid in groups on submerged or floating plants and covered with mucilage. They hatch in about 2 weeks, depending on the water temperature, and a month later the nymphs become adults.

Invisible support

The best time to watch pond skaters is on a sunny day. A close look at an individual reveals that the tip of each leg makes a dent in the water surface. (If the water is shallow the pond skater's shadow can be seen on the bottom, and each leg is surrounded by a round shadow cast by the dent it makes.) The dents are caused by the weight of the pond skater pressing the legs into the water. The insect does not sink, because the force of surface tension counteracts its weight. The water surface is a film of closely linked water molecules that acts like a very fine rubber sheet. Normally we do not notice this film because it is weak, but it is the force that contains a drop of water in a globule on a polished surface and prevents it from spreading out in a pool.

Pond skaters sometimes enter the water completely but can refloat themselves because the undersides of their bodies are covered with a dense pile of hairs that traps air and prevents the body from becoming wet. It has been said that a marine *Halobates* becomes wet and drowns if a drop of rain lands on its back. Yet if this is so, it is surprising that the pond skaters of this genus have not been wiped out by tropical storms.

Gerris lacustris makes a meal of a fly that it has caught on the surface of the water. Sometimes pond skaters gang up to attack prey that an individual cannot overcome alone.

POND SNAIL

The great pond snail has a thin-walled, fragile shell. This species can be found in Europe, northern Asia and North America.

To some people, any kind of freshwater snail is a pond snail, even if it lives more often in rivers. The name pond snail, however, may be restricted, and is for the purpose of this article, to those snails with pointed spires and more or less oval mouths to their shells that make up the family Lymnaeidae. They are the most common freshwater snails in both Europe and North America.

Freshwater snails are of two main kinds. One group, the Prosobranchia, includes the river snails (family Viviparidae) and the valve snails (family Valvatidae), which can close the openings of their shell with a lid, or operculum. The snails of the second group, the Pulmonata, lack an operculum. The Pulmonata are so called because they have a lung, formed by the mantle cavity. The Lymnaeidae belong to this group, as do the bladder snails (family Physidae) and the ramshorns (family Planorbidae), which are also known as trumpet snails and are called wheel snails in North America. Snails of the family Lymnaeidae have a lobe of tissue by the opening of the lung that forms a breathing tube—a kind of natural snorkel.

One of the largest species of the Lymnaeidae is the great pond snail, *Lymnaea stagnalis*, which grows to about 2 inches (5 cm) long, depending largely on the volume of water in which it has lived. At the other extreme, a full-grown dwarf pond snail, *L. truncatula*, may reach only ⅓ inch (1 cm). Unlike those of trumpet and bladder snails, the shells of pond snails are usually coiled to the right. The whorls increase in a clockwise direction as seen from the tip of the spire. In the ear pond snail, *L. auricularia*, the last whorl of the shell is greatly expanded to give the whole an "ear" shape, and this peculiarity is taken to an extreme in the bubble-thin shell of the glutinous snail, *Myxas glutinosa*. This species is unique in that the mantle tissue can spread outside the shell and all but cover it, making the snail look like a little dab of glue. The shells of pond snails are generally a dark or pale horn color and are carried with the spire directed backward. The ear pond snail, however, points its spire to one side.

On the head, pond snails have two nonretractile tentacles, each with an eye at the base. In these respects the pond snails and other freshwater Pulmonata, known collectively as the Basommatophora ("basal-eyed ones") contrast with the garden snail and the other Stylommatophora ("stalk-eyed ones"). The latter have two pairs of tentacles that can be pulled into the body, and the eyes are borne at the tips of the hindmost pair.

Not always in ponds

Pond snails creep along by waves of muscular contractions moving forward over the foot. They sometimes crawl upside down at the surface of the water. Some pond snails, like the great pond snail, crawl to the surface at intervals to renew the air in their lungs. If they cannot reach the surface (for example, when they are under ice) they accelerate their normal skin-breathing. Some species depend entirely on the exchange of gases through the skin and have their lungs permanently filled with water. Pond snails have also been known to survive in ice.

Pond snails live in a variety of freshwater habitats, although with the exception of the ubiquitous wandering pond snail, *L. peregra*, they are not found in very soft or acid water. Some require large bodies of water, but others, like the marsh pond snail, *L. palustris*, and the dwarf pond snail, flourish in ditches or simply on wet mud and can survive long periods of drought.

Rasp their food

Some pond snails live mainly on scumlike algae and other plant material, which is why they are kept in aquaria to keep the glass cleaned of algae. The great pond snail feeds both on animal and plant matter and even kills such animals as newts, small fish and the larvae of water beetles. Some species eat carrion or even resort to cannibalism. All pond snails rasp their food to tiny fragments with their filelike tongue, or radula, which is covered with thousands of teeth.

POND SNAIL	
PHYLUM	**Mollusca**
CLASS	**Gastropoda**
ORDER	**Pulmonata**
SUBORDER	**Basommatophora**
FAMILY	**Lymnaeidae**
GENUS	***Lymnaea, Myxas*; others**
SPECIES	**Many, including dwarf pond snail, *Lymnaea truncatula*; ear pond snail, *L. auricularia*; great pond snail, *L. stagnalis*; North American pond snail, *L. catascopium*; and glutinous snail, *Myxas glutinosa***

LENGTH
Up to ⅓–2 in. (1–5 cm)

DISTINCTIVE FEATURES
Coiled shell with whorls increasing in clockwise direction as seen from tip of spire; 1 pair of antennae with eyes at base

DIET
Plants and algae; some species predatory

BREEDING
Hermaphroditic; number of eggs: 2 to 100; hatching period: 2–3 weeks (*L. truncatula*)

LIFE SPAN
1–2 years

HABITAT
Generally fresh water; some species in muddy ponds or moist soil

DISTRIBUTION
Worldwide

STATUS
Not known

Mating chain reactions

Pond snails are hermaphroditic, each individual having male and female organs opening separately to the exterior. In the dwarf pond snail, a single individual fertilizes itself. Tens of thousands have been reared in the laboratory without cross-fertilization being seen. Usually, though, one individual fertilizes another, one taking the part of a male in every mating and the other taking the part of the female. Sometimes the one acting at that moment as a female acts simultaneously as the male to a third individual, and in this way chains of mating snails may be built up. Reciprocal exchange of sperm between two snails is also possible.

Eggs are laid in gelatinous plates or cylinders attached to a firm support, with 2 to 100 eggs in each mass, according to species and food supply. Development is rapid. Young dwarf pond snails may emerge from the egg mass in 2–3 weeks and mature sexually in a further 3 weeks.

Fluke hosts

The dwarf pond snail is host to the sheep liver flukes. This snail flourishes on damp mud, feeding on the algae on the surface. A connection between liver rot and damp places was known for a long time, but it was the discovery of the role of the snail that opened the door to effective control of the disease. Other water snails are hosts to other kinds of flukes, including human blood flukes. Each species of fluke is to some extent limited to particular snail species. Specifying the snails that support flukes is one of the main problems facing those who study the liver flukes. In Britain and many other countries, only the dwarf pond snail supports liver flukes, but other species with similar habits may take its place elsewhere.

A snail viewed from beneath as it travels across a sheet of glass. In the center of the picture is the base of the animal's foot, with which it moves. The curling, limblike growths to the left of the foot are the snail's tentacles.

POORWILL

The mottled gray-brown plumage of the common poorwill provides the bird with effective camouflage in the woodlands and sagebrush scrub of its native habitat.

THE COMMON POORWILL is a nightjar, about 8 inches (12 cm) long with the drab brown plumage typical of nightjars. Its back is mottled, and the underparts are barred, while the throat is white bordered with black and the short tail is white tipped. The short, wide bill has a large gape and is surrounded by long bristles; the eyes are large. The common poorwill resembles the whippoorwill, *Caprimulgus vociferus*, though the former is smaller and less white in the tail than the latter.

Panting to keep cool

Unlike many nightjars, the common poorwill is not confined to woodlands but is also found on prairies and in arid country where the only cover is scattered sagebrush. It is even found in environments as intemperate as Death Valley in California. It keeps cool at temperatures above 100° F (38° C) by panting, enabling it to lose heat by evaporation from its respiratory passage. The common poorwill has a very low basal metabolic rate (the speed of body processes when at rest). The body does not produce much heat when the bird is inactive, so less strain is placed on heat-controlling mechanisms such as panting.

Poorwills call on and off throughout the night, especially in spring and late summer. Their common name derives from the persistent two-note call. At close quarters a third note is audible and the call can be rendered as *poorwill-low*. Two other North American nightjars are named after their calls: the whippoorwill and the chuck-will's-widow, *C. carolinensis*. The calls are the avian equivalent of the nocturnal buzzing of cicadas and crickets.

Poorwills catch nocturnal insects such as moths, beetles and bugs in flight, scooping them up in the trap formed by the wide open bill and its surrounding bristles. The hard parts of the insects' bodies are ejected as pellets. Poorwills hunt nearer the ground than many other nightjars and sometimes land to search for insects.

Snakelike hiss

Poorwills nest from May to September in the north of their range and from March to August in the south. The two eggs appear white but are actually cream with a pinkish tint. They are laid in a shallow scrape in the earth or on bare rock or shingle, sometimes in the open but more often under a bush. Both parents incubate the eggs, with the male taking more responsibility for the duty in the day. Sitting poorwills are very difficult to find as their mottled brown plumage blends in with their background, but if disturbed they utter a snakelike hiss or raise the wings high above the back, actions that may deter predators.

Poorwills and hibernation

For some time there was much scientific debate over the question of whether birds hibernate. Native Americans of the Hopi tribe have apparently been aware for some time that the poorwill could hibernate, for they call it *holchko*, meaning "the sleeping one," yet there are only two well-documented reports of hibernating poorwills. This may partly be due to the difficulty of finding them. Proof that birds hibernate came only in 1946 when the naturalist E. C. Jaeger discovered a hibernating poorwill in the Chuckawalla Mountains of the Colorado Desert. Jaeger found the bird by chance in a small depression in a granite wall. At first the poorwill showed no sign of life, but as it was repositioned in its depression, it slowly opened and closed one eye. Two days later Jaeger returned to find the poorwill in exactly the same position. He could not detect any heartbeat or breathing and the bird's body temperature was 64° F (18° C), 8° F (4.5° C) lower than normal. Jaeger affixed a numbered band to the poorwill's leg, and for four successive winters he found the bird in the same place. In this way the common poorwill became the first species to show that birds hibernate.

Common poorwills have also been induced to hibernate in captivity. They become torpid when the air temperature drops to 38° F (3.5° C) and the rate of oxygen consumption, an indication of the metabolic rate, drops to one-thirtieth of the normal rate at rest. It has been shown that about ⅓ ounce (9.4 g) of fat is sufficient to sustain a hibernating poorwill for 100 days, long enough to tide it over during the winter dearth of insects.

Since the late 1940s, investigations into the possible use of torpor by nightjars have been undertaken on both wild and captive birds of several species, including the common poorwill, the Eurasian nightjar (*C. europaeus*), the common nighthawk (*Chordeiles minor*), the whippoorwill and the spotted nightjar (*Eurostopodus argus*). However, it seems the common poorwill is the only nightjar regularly to use this capacity.

COMMON POORWILL

CLASS	**Aves**
ORDER	**Caprimulgiformes**
FAMILY	**Caprimulgidae**
GENUS AND SPECIES	***Phalaenoptilus nuttalli***

LENGTH
Head to tail: 7¼–8⅓ in. (18–21 cm)

DISTINCTIVE FEATURES
Short, broad bill with huge gape; large eyes; gray-brown or gray-white upperparts, with brown speckles and bars; dark brown chin and throat; white lower throat; rest of underparts pale grayish brown

DIET
Beetles, moths, cicadas, flying ants and flies

BREEDING
Age at first breeding: 1 year; breeding season: March–September; number of eggs: 2; incubation period: 19–24 days; fledging period: 20–23 days; breeding interval: 1 year

LIFE SPAN
Not known

HABITAT
Arid or semiarid country, usually at altitudes of 1,650–3,300 ft. (500–1,000 m)

DISTRIBUTION
Southwest Canada, western U.S. and northern Mexico

STATUS
Locally common in southwest of range; uncommon elsewhere

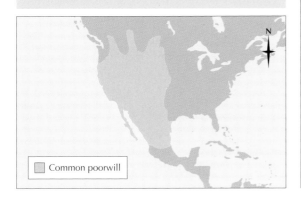

Common poorwill

In common with other poorwills, the ocellated poorwill, Nyctiphrynus ocellatus, is nocturnal. It roosts by day on rocks or branches, becoming active at dusk.

Index

Page numbers in *italics* refer to picture captions.
Index entries in **bold** refer to guidepost or biome and habitat articles.

Page numbers in *italics* refer to picture captions. Index entries in **bold** refer to guidepost or biome and habitat articles.

Page numbers in *italics* refer to picture captions. Index entries in **bold** refer to guidepost or biome and habitat articles.